GREG PAPANDREW

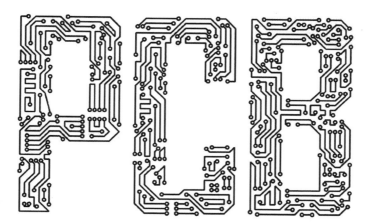

PCB

BASICS FOR BUYERS

A QUICK GUIDE TO THE
PRINTED CIRCUIT BOARD INDUSTRY

Better Board Buying™

PCB Basics for Buyers: A Quick Guide to the Printed Circuit Board Industry

By Greg Papandrew
Better Board Buying LLC
BoardBuying.com

Published by JMP Publishing LLC
204 4th Ave., #908
Indian Rocks Beach, FL 33785

Paperback ISBN: 978-0-9839391-9-1

Table of Contents

Foreword

Electronics is the largest industry in the world. Much of the attention revolves around the tools that improve our everyday lives—products that range from the mundane and nearly transparent such as the controls for refrigerators, to the Uber-hyped game-changers like the tablet PCs and the iPhone.

And at the core of every one of these systems is the printed circuit board.

The printed circuit, whose origins date to the turn of the 20th century (although the history of electroplating is even older—by another 100 years), is the foundation upon which all electronics are built. Those ubiquitous green boards, visible through the air vents of audio-visual equipment or when changing the battery on your phone, are the spine through which a vast network of data is circulated at the speed of light, allowing you to make a phone call, or listen to a song, or surf the internet, or reminding you to wake up.

How printed circuit boards, or PCBs, are produced is a complex process involving dozens of steps, whereby epoxy-glass laminate is layered with sheets of copper and heated, etched, immersed in a "bath" with an electrical current applied, coated, tested, and cleaned more times along the way than you can count.

What PCB veteran Greg Papandrew has tried to do is explain that sophisticated process using a proven series of basic concepts, illustrations, and examples, infused with a glossary of technical terms because, like any industry, PCBs come with their own set of jargon.

I think Greg has succeeded.

Happy reading!

Mike Buetow

Editor-in-Chief, *Printed Circuit Design and Fabrication*

Introduction

Whether you are a novice in the printed circuit board assembly industry, or an old hand who sometimes needs a refresher, *PCB Basics for Buyers* is for you. It gives you a straightforward understanding of the board business and what it takes to make a PCB.

The book contains detailed images of the manufacturing process, information on specs and standards, tips for creating successful PCB partnerships, and a glossary of industry terms. It's a great resource and training tool for your entire procurement team.

The pressures on the board industry are enormous, from international trade tensions to high capital equipment costs, and from increased environmental compliance standards to workforce challenges. That makes it even more important that PCB buyers—who play a key role in keeping the circuit board supply chain rolling—know what they're doing.

And yet, most PCB buyers get no formal training. Remarkably, even ISO-certified EMS and OEM companies simply don't have the resources to provide formal instruction in buying a custom-made commodity that is the foundation of their assembly process.

My company, Better Board Buying (B3), is on a mission to change that. Our dedicated training program gives buyers the tools they need to do their jobs better, offering the most up-to-date information about PCB purchasing. In the past, board buyers have saved 10 to 25 percent by employing our methods.

B3 teaches buyers how to evaluate the strengths and weaknesses of both off-shore and domestic manufacturing sources. We also show them how to leverage buying power to get the best prices, and how to expertly navigate sourcing issues based on order size and technology type. And that's just for starters. In short, we build better buyers by turning them into PCB problem solvers.

I hope this book is helpful to you in navigating the often-stormy waters of this $60 billion business.

Interested in learning more about Better Board Buying? Contact me at greg@boardbuying.com. Or visit BoardBuying.com.

Greg Papandrew

Chapter 1

What is a PCB?

A PRINTED CIRCUIT board (PCB) provides a surface for mounting electronic components and electrically connects these components. You may have also heard it called a circuit board, bare board, printed wiring board, or circuit card. No matter what it's called, PCBs are used in all but the simplest electronic products. Most things that use electricity have at least one circuit board that makes them run.

Each PCB is a custom device, usually designed by one company for a specific product, yet often manufactured by another. On top of that, it's often assembled by a third company. It's generally the last part of an electronic product to be designed, but the first part needed to start assembly.

The manufacturing sequence for a circuit board consists of more than 100 individual steps, involves a variety of raw materials, and employs mechanical, chemical, electrical, and photoimaging processes.

Introduced in the early 1900s, PCBs were once merely single-sided products, meaning they only had circuitry on one side. Single-sided boards are still in use today, mostly for consumer electronics; they are manufactured in great volume and are the least expensive to produce.

In the 1970s, the industry developed processes for plating copper on the walls of the drilled holes in circuit boards, allowing top and bottom circuitry to be electrically connected. These double-sided boards quickly became the standard.

As electronic components became ever more complex, the multilayer board was developed. Multilayers, which have several circuitry layers sandwiched together, are used in a wide variety of sophisticated electronic components and are the more expensive PCBs to produce.

PCBs are built into many kinds of equipment, ranging from smartphones to cars to powerful computers. They can be small enough to fit inside hearing aids, or large enough to power heavy industrial machines. They can be rigid or flexible, molded, or even three-dimensional.

The board part of a PCB is usually made of fiberglass, a material that does not conduct electricity. Typically, copper is etched on the surface of the board or, if multilayer, inside between layers of fiberglass. The conductive pattern that is etched into the board lets electricity travel from one component to another in circuits, ensuring the electric current goes only where it is wanted. The components are then attached to the board with a metal (a process called soldering) to conduct the electricity and provide a strong adhesive for the electronic components.

Because there are so many different uses for PCBs, board manufacturers generally do not design the boards they produce. All of the specifications of an individual board—such as shape, mechanical and electrical properties, surface finish, material composition—are provided by designers. These designers typically work for an original equipment manufacturer (OEM), or a design service bureau, or a contract electronic manufacturer (CEM).

The terms *fabricator* and *manufacturer* are used interchangeably because they are the same thing.

Depending on where the circuits are placed, PCBs fall into three categories:

- **Single-sided Boards:** Boards with circuits on one side are *singled-sided.*

- **Double-sided Boards:** Boards with circuitry on both sides are *double-sided.*

- **Multilayer Boards:** Boards with layers of circuits inside the base material are called *multilayers* or *multilayer boards* (MLB). Multilayers are further defined by the number of layers they have, ranging from as few as three to more than 50.

What It Takes to Make a PCB

Three raw materials go into the composition of a printed circuit board:

- **Laminate:** *Laminate* is woven-glass material that is reinforced with resin. The reinforcing material is used as a base to hold resins and give the board stability. Epoxy glass in the middle serves as insulation and provides structural strength for mounting components. Copper foil on the outside is the conductive medium through which electrical currents travel.

- **Prepreg:** *Prepreg* is a sheet of woven-glass reinforcement impregnated with a resin that is not yet fully cured (also known as *B-stage resin*). See Chapter 3 for more information.

- **Metal foil:** Copper is the most commonly used metal foil.

Putting A Board Together

The fabrication process starts with the laminate (also called the *dielectric* or *substrate*) as the primary raw material. *Traces* (also called *conductors* or *circuits*) create electrical interconnections on the laminate.

Traces are made by selectively removing portions of the copper foil. Electrical current is carried by copper deposited on the walls of holes drilled in the boards. This connects the top surface circuitry to the bottom, as well as to the layers of circuitry inside the laminate.

Multilayer boards are created using a combination of laminate, prepreg, and copper foils. As layers are stacked, they are put into an oven and pressed.

The majority of boards manufactured today are 0.062 inches thick. But some boards are almost half an inch thick, while others are as thin as a sheet of paper. Approximately 90 percent of PCBs are *rigid* boards, meaning they are made using a woven glass material that is stiff or rigid enough to support a variety of components. Rigid boards are used in most products that require an electrical interconnection.

Flexible circuits, on the other hand, are very thin PCBs capable of bending. And rigid-flex boards are a hybrid product made from both flexible and rigid technologies. These PCBs are used predominantly in high-tech applications and devices for the military, aerospace, and medical industries, where space is at a premium and the PCBs can be molded or bent to fit into a very small package. Figure 1-1 shows the parts of a PCB.

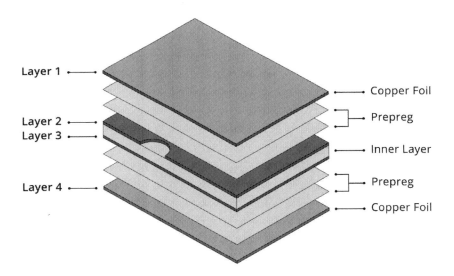

Figure 1-1: Stackup and multilayer construction

Mounting the Components

Most of today's circuit boards have components mounted to them. These components activate and monitor the electrical current that passes through the board.

Components are mounted to PCBs in either of two ways:

- **Plated Through-hole Technology (PTH)**: In this method, component legs (called *leads*) are inserted through holes in the PCB and attached to *lands* on the opposite side of the board (for more information on lands, see the sidebar "Land Lesson"). A PTH is a drilled hole in a PCB that has been plated with copper metal, allowing a connection between the conductive patterns. Holes drilled large enough to accept the lead of a component, pin, or wire to be soldered to the PCB are called *component holes*. A hole too small for a component, but which allows for a layer-to-layer connection, is called a *via hole* or *via*.

- **Surface-mount Technology (SMT)**: In this method of mounting components, component leads (also known as *surface-mount devices* or *SMDs*) are soldered directly to pads on the surface of the board.

Of the two assembly techniques, SMT is more commonly used.

Figure 1-2 shows the differences between PTH and SMT.

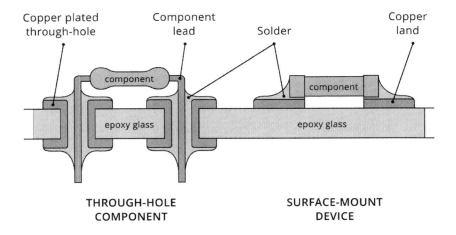

Figure 1-2: Plated through-hole and surface-mount technology

Land Lesson

Many people use the term *pad* when the correct term is *land pattern*. A land is any metallic surface of a PCB that is not covered by solder mask and allows for the soldering of components, pins, or wire. The annular ring around a component hole is also known as a land, even though most people call it a *pad*.

Chapter 2

PCB Specifications & Standards

IN THE PRINTED circuit board industry, a specification is a document that defines and quantifies the mechanical, chemical, and electrical properties of a PCB.

Over the years, specifications have been written to define raw materials, appearance, workmanship standards, and methods of testing, as well as for detailing the data format needed for describing a PCB electronically.

Standards—guidelines that are agreed upon by a range of industry professionals and trade associations—are maintained through specifications that establish uniform quality benchmarks for finished products.

Specifications and standards serve to set criteria and ensure uniform quality for finished products. Private, public, and government agencies have collaborated to create the various specifying documents.

Who Sets the Standards?

Three organizations are recognized as the primary sources of initiation, implementation, and control of PCB documents in the United States:

IPC

IPC (formerly known as The Institute for Interconnecting and Packaging Electronic Circuits) is an international trade association of fabricators, assemblers, original equipment manufacturers (OEMs), and suppliers that was originally founded in 1957 as the Institute of Printed Circuits.

In the years since, the IPC has taken the lead in generating new standards for raw materials, design, qualification and performance, processing, testing, and overall acceptability of PCBs.

For more information on current IPC specifications and standards, as well as a graphic representation of how all the guidelines work together in the PCB manufacturing process, visit IPC.org.

Underwriters Laboratories

Underwriters Laboratories (UL), an independent laboratory, performs site inspections and laboratory testing, checks documentation, equipment, calibration, and test records, and verifies that samples and tests are being performed properly.

A recognized UL symbol—a logo or company abbreviation with a dash and number such as XYZ-3—indicates that the PCB made by a particular fabricator from a particular laminate and subjected to a particular sequence of processes meets certain test criteria. UL requires a fabricator's recognized mark to appear on its finished boards.

Proof of UL certification, the published "E" number for a particular PCB manufacturing facility, can be easily obtained by asking the supplier or by downloading a copy found on the certification database at UL.com.

UL also tests and issues a flammability rating for each laminate/process combination. These ratings—94HB, 94V-2, 94V-1, 94V-0—indicate how well a particular combination supports combustion. A 94V-0 rating means virtually no support of combustion, and 94HB indicates no flame-retardant properties.

In practice, material manufacturers obtain UL flammability ratings for their materials, whereas a PCB manufacturer obtains a UL symbol for different layer counts, various materials, and processes used.

U.S. Department of Defense (DoD)

The Defense Supply Center Columbus (DSCC), located in Columbus, OH, is the controlling government agency that defines the manufacture of *mil spec* PCBs used in military and other government-related applications. DSCC maintains a list of PCB fabricators that have passed an audit and allows only certified manufacturers to supply the DoD.

The DoD, through DSCC, has issued a series of specifications that govern any PCB to be used in military products. These specifications include:

- MIL-PRF-31032, "Printed Circuit Board/Printed Wiring Board, General Specification For".

- MIL-PRF-55110, "Performance Specification, Printed Wiring Board, Rigid, General Specification For".

- MIL-PRF-50884, "Printed Wiring Board, Flexible or Rigidflex, General Specification For."

Any PCB fabrication print that calls out these specifications must be produced by a government-approved PCB manufacturer. For more information, visit the Land and Maritime section of the Defense Logistics Agency website.

How to Contact:

IPC	Defense Supply	Underwriters Laboratories
3000 Lakeside Dr.,	**Center**	333 Pfingsten Road
Ste 105N	PO Box 3990	Northbrook, IL 60062-2096
Brannockburn, IL	Columbus, OH	(847) 272-8800
60015	43218-3990	ul.com
(847) 615-7100	(877) 352-2255	
ipc.org	dla.mil/	
	LandandMaritime	

Directing Performance through Documents

The standards organizations regularly produce *performance* documents—published specifications and guidelines—and other publications that govern practices in the PCB industry. The better you get to know the industry, the more you're likely to hear about the following types of documents:

- **IPC Documents:** The IPC issues documents relating to materials, processes, testing, component mounting, phototool generation, and other topics.

- **Customer-supplied Documents:** The printed circuit board user often creates in-house documentation or internal specifications that state that company's particular needs. These in-house specs may reference IPC documents, military standards, and UL ratings in addition to particular workmanship and acceptability standards, as well as other criteria that the company feels are critical. The fabrication notes of the customer-supplied print should reference these requirements.

- **UL Directory:** UL publishes the *Underwriters Laboratories Recognized Component Directory,* which lists all approved fabricators and their recognized UL symbols, as well as summaries of the related test data.

Spec Check: IPC-A-600 and IPC-6012

IPC provides two main specifications that pertain to manufacturing rigid PCBs:

- **IPC -A-600: Acceptability of Printed Boards:** This workmanship standard describes preferred, acceptable, and non-conforming conditions for PCBs.

- **IPC-6012: Qualification and Performance Specifications for Rigid Printed Boards:** This standard covers the testing, qualification, and documentation required to prove the PCB meets or exceeds the classification specified by the customer.

These two standards recognize that every PCB has an intended end-use, so they place PCBs in three categories—Class 1, Class 2, and Class 3—which reflect the complexity of the PCB, as well as its reliability, performance, and required verification (testing).

Class 3 requires the highest workmanship standard, the most testing, and the longest retention of documentation, so it costs the most.

As a PCB buyer, it's crucial that you fully understand the fabrication notes supplied by the designer. Sometimes buyers erroneously assume a higher level of inspection is required on an order, unnecessarily driving up costs. Unless the board is intended for a known military, DoD, medical, or aerospace purpose (or any application that needs extremely high reliability), IPC-6012 Class 3 certification is probably not required.

The majority of PCBs purchased require only the observable IPC-A-600 Class 2 inspection, along with electrical tests and the associated first article paperwork. Find out exactly what you need before sending out a quote, and you will save time and money.

This book covers only information relevant to *rigid* circuit boards, which make up about 90 percent of the PCB industry. *Flexible* and *rigidflex* printed circuit boards have their own specifications.

IPC-4101: Specification for Base Materials for Rigid and Multilayer Printed Boards

IPC-4101 covers the requirements for laminate or prepreg for rigid or multilayer PCBs. The specification defines the electrical, mechanical, chemical, and environmental requirements that the various combinations of reinforcement and resin (laminate) must meet.

Individual specification sheets (sometimes called *slash sheets*) represent the variety of materials available. Each sheet outlines requirements for both laminate and prepreg for each product grade. The specification sheets are organized by reinforcement type, resin system and/or construction. They are identified by specification sheet number for ordering purposes.

Chapter 3

Designing and Defining PCBs

PCB DESIGN BEGINS when an electrical engineer chooses the components required to perform the functions of the end product and then determines the best way to connect those components electrically.

Whether a board is simple or complex, the design gives the fabricator a great deal of manufacturing information, including the PCB dimensions, hole sizes and locations, and overall mechanical definition; it may also incorporate notes referencing type of material, workmanship standards, UL requirements, and solder mask and test requirements, among other things.

Let's talk about the process of designing a PCB, from setting the specs to generating tooling data.

CAD: Creating Hardware with Software

While PCB designs were once played out manually on mylar film, they are now created using computer-aided design (CAD) software (also known as *CAD tools*) that is made specifically for PCB layout.

Since the 1990s, the industry has followed IPC specifications that provide a standard for data formats to be used for transmitting information between a printed circuit board designer and a manufacturing facility.

Automating the Blueprint

The CAD software automatically determines the path (route) for the conductors, reducing the manual labor required.

Another option for CAD programs is *schematic capture*, in which the designer enters a digital drawing of the electrical paths into the software program, and the program designs the board layout.

Most CAD vendors provide a library of shapes and sizes of available components. These shapes and sizes are known as *outlines*. The area where these components come into contact with the board is called the *footprint*.

Selecting Materials

Because a variety of copper thicknesses, epoxy properties, and types of glass weave are available, the designer must define the desired combination. One important consideration is the makeup of the *prepreg*—the agent that bonds the layers of a multilayer PCB. Prepreg (or *B-stage resin*) is a glass cloth that has been impregnated with epoxy and then partially cured. This cloth is available in a variety of glass weaves and epoxy compositions. Selecting the right materials is crucial to ensure the PCB and its circuitry remain reliable throughout the life of the product.

There are three basic properties—thermal, electrical, and chemical—that board designers consider when choosing material for PCB construction. Their selections will vary depending upon the desired application, and these parameters have a direct effect on the bonding (lamination) process, so the selection of the proper material is critical.

A PCB material's thermal properties govern its ability to withstand extreme temperatures while retaining its characteristics. For example, a circuit board will be baked to drive out moisture and then subjected to reflow temperatures during assembly.

Once it is assembled, a PCB can experience temperature fluctuations, and its components can also generate heat. Depending on the thermal properties of the materials used, the PCB could soften, and in extreme cases, the layers can separate, an effect known as *delamination*. Choosing a material with suitable thermal characteristics and the ability to remove heat (known as *thermal conductivity*) is essential.

The electrical properties inherent in a board material also influence its selection. For some high-voltage applications, low electrical leakage in the PCB material is essential. And the way the board will ultimately be used may require high signal integrity. During the design phase of the PCB, attributes such as the height of the trace, the distance to other copper features, and the dielectric constant of the material (to name a few) can be manipulated to reach the desired level of signal integrity.

A board may also require *controlled impedance*, which, put simply, is the ability to control the speed of an electrical circuit. Because the traces of a circuit board can no longer be regarded as simple conductors—as components have become smaller, faster, and more complicated—some specific functions of these components need to take place before others. Controlled impedance, which is measured in Ohms, makes this possible. Controlled impedance boards do require the use of time domain reflectometry (TDR) testing and specially designed coupons for testing and validation purposes. For more information on controlled impedance, see Chapter 5.

For safety, it is crucial that the chemical properties of the PCB material follow safety standards, including being flame-retardant and having a low value for moisture absorption. PCBs must also be able to withstand chemical cleaning after assembly.

Specifying Metallic Finishes

The designer's manufacturing information includes notes on the type and method of finish to be used. Copper is used as the conductor in most PCBs in production, and if left unprotected, will tarnish and prevent the proper soldering of components to the bare board. A metal that doesn't tarnish or is slow to tarnish is applied to protect the exposed copper surfaces of the PCB. Metallic surface finishes vary in price, shelf life, reliability, and assembly processing.

Most surface finishes are considered *SMOBC* (Solder Mask Over Bare Copper). Solder mask is generally an epoxy-based, thermosetting resin applied by a variety of methods (see Chapter 4). It covers and insulates all the traces and some of the lands, except those to which components will be soldered.

The advent of lead-free legislation and the Restriction of Hazardous Substances Directive (RoHS) has given the industry some challenges. Although each finish has its own benefits, in most cases, the process, product, or environment will dictate the surface finish that is best suited for the application. The end user, designer, and assembler should work closely with the PCB supplier to select the best finish for the specific product design.

Some of the more prominent surface finishes are:

- **Hot Air Solder Leveling (HASL):** In this method, the PCB is dipped into a bath of molten solder (tin-lead) so that all exposed copper surfaces are covered by solder. Excess solder is blown level using compressed air.

- **Electroless Immersion Gold (ENIG):** This is one of the best and most common RoHS finishing methods for printed circuit boards. ENIG chemistries offer many strengths, including excellent wetting, coplanarity, surface oxidation, and long shelf life.

- **Immersion Silver:** Immersion Silver is deposited directly on the copper surface by a chemical displacement reaction and provides a RoHS-compliant finish for PCBs.

- **Organic Solderability Preservative (OSP):** OSP is a RoHS-compatible, water-based organic compound that selectively bonds to copper and provides an organic/metallic layer that protects the copper during soldering.

- **Lead-Free HASL:** This is a RoHs alternative to regular HASL where the physical application is the same, but the molten bath is free of lead and uses a nickel-modified alloy instead to give similar results.

Creating the Stackup Sheet

As the final step, the designer creates a *stackup sheet* (see Figure 3-1) that provides an overall picture of the board. This sheet is a computer-generated mechanical drawing of the conductive and insulation materials to be used in the board and the order in which they should be placed, or *built up.*

Stackups

The *stackup* is the specific call-out of material thickness and copper weights required to produce a particular multilayered PCB. The description of the stackup can be found on the board's fabrication drawing — a drawn cross-section depicting the material and copper thickness of each layer involved. The stackup is how the designer tells the manufacturer the way to properly build the PCB so that it will work as designed.

Circuit board materials available for multilayer construction are primarily of two types: *core material* that has copper laminated to both sides (cured laminate) and is usually 0.031 inches or less thick; and *prepreg* material (uncured, resin-treated glass) that is used to fill the spacing between cores and outer-layer foils and is used to bond all layers together. *Copper foil* of various weights (quarter ounce or greater) is available as well for outer-layer construction. During the manufacturing process, this laying, or stacking up, of materials involved is called *creating a multilayer book.*

Using a six-layer PCB as an example, the book is created this way: Copper foil is laid down as Layer 6, or the first "page" of the PCB, followed by a layer—or layers, as required—of prepreg. The core material that contains Layers 4 and 5 —or in this case "pages" 4 and 5—is applied on top, followed by an additional layer or layers of prepreg. "Pages 2 and 3" are added with the additional core material that will contain Layers 2 and 3. Then, it is covered with the final page, Layer 1, and the book is complete.

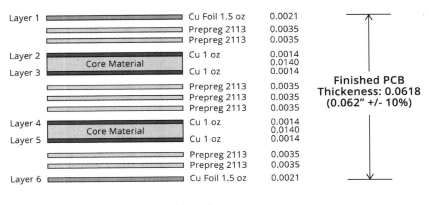

STACK UP

Figure 3-1: An example of a stackup sheet

Depending on the manufacturing equipment, several books may be stacked and sent to the press for the lamination process. The press can be either vacuum, hydraulic, or both, where the removal of entrapped air, even pressure and temperature are applied to ensure the baking and curing of the materials are done correctly to become one solid board.

The stackup sheet shows the complete stackup for all types of printed circuits. For a multilayer board, it also defines the materials and manufacturing method to use for each layer.

The designer may specifically define, or ask the PCB manufacturer to define, the buildup to control electrical properties such as impedance and current-carrying capacity.

Generating Manufacturing Data

Electronic data describes the location of holes and conductors in terms of an X- and Y-coordinate system. This data, called *CAD data*, can be used to generate photographic images of the PCB and to produce other manufacturing tooling items. Thus, the CAD data can be used in many areas of the factory, saving time, and improving accuracy and quality.

Also incorporated into CAD data are notes that reference the following:

- ✓ Workmanship standards (internal or industry specifications).

- ✓ Underwriters Laboratory (UL) requirements.

- ✓ Test requirements (see "Step 3: Programming Inspection and Test Data" in this chapter).

- ✓ Any special requirements (customer-specific requirements).

Transferring the Design to the Fabricator

The board layout data and manufacturing information are stored on the design computer and typically transmitted to the fabricator via the internet. The two most-used data-transfer formats are a basic one called *Gerber* and its derivative, *Gerber RS-274X.* Various companies and trade associations have tried through the years to develop and introduce superior formats; ODB++, GenCAM, and PIC-D-350 also are in use today.

ODB++ is a proprietary CAD-to-CAM data exchange format used in the design and manufacture of electronic devices. Its purpose is to exchange printed circuit board design information between design and manufacturing, and between design tools from different EDA/ECAD vendors.

Converting the Design for Tooling

The manufacturing process begins with the manipulation of the electronic data received from the designer, using special programs known as *computer-aided manufacturing (CAM)* software. This software permits the fabricator to perform panelization, and to generate drilling and router programs. It also generates data to be used in electrical test and optical inspection and transfers the design data to laser photoplotters to generate photographic images of the PCB on film.

Thus, the work performed on a CAM system is often referred to as the *front end*, also known as *engineering*. The front-end system converts customer information to manufacturing *tooling*, which is the general name given to items necessary for PCB manufacturing such as films, drilling and routing data, and test fixtures.

Step 1: Performing Panelization and Phototooling

The first step in the tooling cycle is *panelization*, a process is which multiple images of the PCB are generated to fill the entire surface of the panel of laminate economically. This process is performed on software that uses the *step-and-repeat* method—combining successive exposures of a single PCB image into a multiple-image production master—to create the desired configuration.

Next, auxiliary features—layer numbers, UL symbols, borders, and *test coupons* (a tiny mockup PCB used to verify the panel is correct)—are added to the panel.

Panelization for each layer of circuitry, plus the features required for solder mask and component legends, is prepared in the same way. The conduct trace and feature widths can be controlled to compensate for etching and processing variations. The CAM software also electronically *registers*, or aligns, the various layers.

The panelization data is stored electronically until needed. Then it's transferred to the *photoplotter*—a machine that draws the panelized image with laser light through a process known as *imaging*. The final image is drawn on photographic film.

This film is the master *phototooling*, or production, master. Some fabricators use the master directly in production; others use it to produce more copies that are in turn used to phototool large volumes of panels.

Laser direct imaging (LDI) is another method of applying or exposing the PCB design to the manufacturing panel by directly imaging the circuitry to the manufacturing panel using a powerful laser. LDI is gaining popularity in the industry for its ability to maintain artwork alignment to meet the increasing practice of using smaller and condensed electronic packaging. Another big plus is that money is no longer spent on films (phototooling) and the costs associated with storing and cataloguing them.

Step 2: Generating Drill and Routing Data

The next step in the tooling cycle is generating the numerically controlled (NC) drill and routing data for transmission to the drilling department.

Drill data tells the computer-controlled drilling machine the locations and sizes of the holes in the PCB. It includes the machine commands to change drill bits, vary spindle feed (RPM), and control the up and down rates (*up feed* and *down feed*) of the spindles. A computer network downloads the data directly to the machines or to special data handlers, thus eliminating handling and storage of drill information.

Routing data, generated the same way as drill data, instructs the computer-controlled router to form the finished board dimensions. Again, direct downloading from the CAM workstation to the routers or to computer networks ensures prompt, correct data transfer.

Step 3: Programming Inspection and Test Data

Before the advent of small, powerful PCs, human workers inspected PCBs for flaws visually, using a magnification device. As the width of the conductors and the spaces between them decreased, however, manufacturers adopted special technology to find breaks or shorts in the narrower conductors, called *fine lines*. This technology is called *automated optical inspection (AOI)*.

Fabricators program the AOI memory by using a *golden board* (also called a *known-good board)* or CAD data as a reference. The AOI machine scans the board under inspection and compares it with the golden board. In the case of a mismatch, the machine identifies the location, enabling the fabricator to find the defect, which can be one-tenth the width of a human hair.

With the establishment of improved data-handling software and the creation of data formats that permit machines to communicate, many manufacturers prefer to generate AOI data from the basic Gerber data supplied by the customer. This method eliminates the need to rely on a golden board and significantly reduces the chance of producing defective circuit boards.

The bare board is electrically tested to ensure it reflects the original design as intended. The test is typically a series of continuity and isolation tests. There are basically two methods for testing:

- ✓ **Golden Board Testing:** This method is primarily used when no Gerber or electronic data is available to generate artwork used in the manufacture of the PCB. Legacy product that is usually through-hole, simple 2-, 4-, and even 6-layer boards are still tested in this fashion. This is a self-learn test when a (hopefully) known good board—the golden board—is used as a reference.

- ✓ **Netlist Testing:** This method can be used when Gerber data is available. It is performed when all the *nets* (a string of points along a circuit) can be extracted from the information supplied. Where golden board testing is a comparison test, the netlist tests all points, ensuring a high confidence of electrical integrity. This true electrical test of point-to-point is just a step away from the actual schematic drawing of the PCB.

Chapter 4

Manufacturing Double-sided PCBs

THE VAST MAJORITY of PCBs produced today are multilayer boards. But to understand how those boards are made, you first need to understand how double-sided PCBS are manufactured.

NOTE: A single-sided board is processed like a double-sided board, except that the fabricator starts with raw material clad with copper on one side only.

Planning and Preproduction

Before manufacturing begins, the fabricator reviews the CAD data and other information sent by the customer. (As discussed in Chapter 3, CAD data includes films, mechanical drawings, and specifications.) Then the fabricator decides on the most effective way to manufacture the PCB, based on the fol-

✓   **Number of boards "up" per panel:** This figure represents the quantity of boards arrayed on a particular panel size (see the following item).

✓   **Panel size:** Panel size is determined by the best manufacturing process of a particular

Let's Build It!

board house where multiple images of the same PCB design are placed. Panel size varies, depending on the technology or volume of the order and the processing capability of the PCB manufacturer. The most common sizes are 18 x 24 inches, 16 x 18 inches, 12 x 18 inches, and 9 x 12 inches.

✓ **Features and information to be added during panelization:** Items such as UL symbols, test coupons, layer numbers, and borders are selected at this time.

✓ **Layer materials:** Often, the designer selects the materials for individual layers, but if the composition isn't critical, the fabricator makes this decision.

✓ **Drilled hole sizes:** Because subsequent operations deposit copper in the holes, the holes must be drilled slightly larger to accommodate the finished size.

✓ **Tooling holes or target locations:** Fabricators have tooling systems that define the locations of tooling features and match their equipment.

The planning or engineering department summarizes these decisions in a document called the *traveler* (also known as a *production* or *shop traveler, routing sheet, job order, or production order*). The traveler instructs the factory how to process the board and includes any other information the factory needs to produce that particular PCB.

The Manufacturing Process

The following section describes the steps involved in producing a double-sided board with solder mask over bare copper (SMOBC), plated through-holes (PTH), and gold-plated contacts and component legend.

The figures in this section depict some of the physical changes that a double-sided PCB undergoes during the production cycle. They do not illustrate the cleaning, rinsing, drying, or baking operations involved, however, and they don't show any processes following the solder-coating operation. The focus is on what happens to the copper foil on the board's surface.

The PCB manufacturing sequence consists of 75 to 125 separate operations, and no two fabricators perform them exactly the same way. The differences in processing methods are due to the wide variety of equipment, materials, and proprietary chemistries available.

The sizes of production facilities and the specific preferences of their managers also lead to variations. Here, we focus on the predominant manufacturing methods, but we also mention some common alternative processes as well.

Step 1: Preparing Materials

Using the information on the traveler—including the numbers and sizes of the panels, as well as any special instructions—the fabricator prepares the materials necessary to process the order.

The majority of PCBs start with copper-clad epoxy glass as the raw material (see Figure 4-1). The traces (circuitry) are formed by applying copper plating to the holes and circuitry, and then etching the base copper away, thereby selectively removing the unwanted copper. This approach is known as *additive/subtractive processing*.

Shearing laminate to the proper size at the fabrication plant can produce copper slivers and shreds of epoxy glass that can cause defects at later stages. Therefore, many fabricators now request that laminate suppliers send the laminate precut to panel size. The same is true of the entry material and the backup board (see the following section).

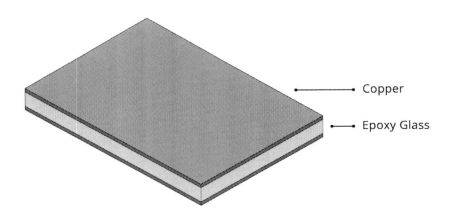

Copper

Epoxy Glass

Figure 4-1: The Starting Point

Step 2: Stacking and Pinning

The copper-clad panels are stacked (most often three panels to a stack), along with the following elements:

- **Entry material:** *Entry material*, which is placed on top of the panel stack, is made of phenolic sheet, aluminum foil, or paper, and ranges in thickness from 0.005 to 0.010 inches. The entry material serves a couple of purposes:

  1. It gives the drill bit something soft to enter before hitting the copper-clad panel, thereby improving accuracy, and preventing *burrs* (little spurs of copper created where the drill enters the panel).

  2. The entry material prevents the pounding foot on the drill machine from denting the copper panel.

- **Backup board:** The *backup board*, which is placed below the panel stack, is a sheet that prevents burrs from forming on the bottom panel and protects the drill table. The backup board is composed of phenolic sheet, paper composite, or aluminum foil-clad fiber composite, and most often is 0.093 inches thick.

The stacks are then pinned together by drilling two holes (usually 1/8-inch -inch or 3/16-inch in diameter) in opposite ends of the panels and pressing steel dowel pins through the holes. The dowel pins (usually 1/2-inch-long) protrude from the stack bottom by approximately 1/4- inch, and eventually locate and hold the stack on the drilling machine table (see Chapter 5, Figure 5.1).

In typical production, holes can be drilled in copper-clad epoxy glass to a depth of five to six times the diameter of the drill without fear of drill breakage or creation of rough holes. The ratio of thickness (or depth) to the smallest hole diameter in a board is called *aspect ratio*. A drill with a diameter of 0.032 inches can safely drill a stack depth of 0.160 to 0.192 inches. With 0.059-inch-thick material, the measurement equals a stack of three panels.

Step 3: Drilling

The stacks—entry material, panels, and backup board—are positioned to the drill using the dowel pins previously installed at the known tooling locations. The dowel pins mate with bushings in the drill table so that the stacks are dimensionally located to the drill *spindles*—devices that hold and turn the drill bit. Today's machines typically have multiple spindles.

Drill data—the X- and Y-coordinate information that defines hole position, drilled hole size, spindle speed, drill-bit feed, and retraction rates—is sent to the drill automatically by the Gerber software panel, as discussed in Chapter 3.

See Figure 4-2 for a look at the PCB after drilling.

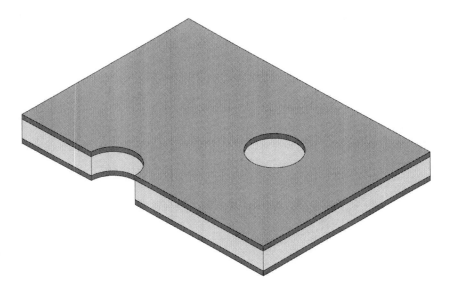

Figure 4-2: Hole size and locations are determined by drill data defined in the shop traveler.

Step 4: Deburring

As drilling processes improve, fabricators are producing virtually burr-free holes. Still, most fabricators process drilled panels through a deburring machine. The panels pass through brushes or abrasive wheels that mechanically remove any copper burrs at the rims of the holes. Deburring also removes any fingerprints and oxides, and creates a smooth, shiny surface.

Step 5: Electroless Copper Plating

After drilling and deburring, the panels are unstacked, placed on racks, and processed through a series of chemical baths. The chemicals remove any organic contaminants and clean the copper. They also sensitize the epoxy glass on the walls of the drilled holes so that the glass can receive a thin coating of copper.

This copper coating alone isn't sufficiently thick to carry the electrical load, but it provides a metalized base on which additional copper can be electrolytically deposited. Electroless copper depositions range in thickness from about 50 micro-inches to 150 micro-inches (See Figure 4-3).

Several environmentally friendly processes introduced in the 1990s have supplanted electroless copper as the predominant process in certain regions. *Direct metallization* uses environmentally friendly agents called *reducers* to deposit a thin layer of copper palladium on the hole walls. The *carbon/graphite method* chemically bonds a layer of conductive carbon or graphite to the hole wall. Both offer enough conductivity to plate the electrolytic copper to the desired wall thickness.

Electrodeposited Copper

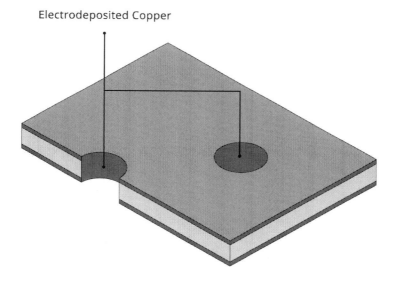

Figure 4-3: A thin layer of copper is deposited on all surfaces, including the walls of the drilled holes.

Step 6: Imaging

In *imaging*, a negative-image circuitry pattern is transferred to the PCB panel. First, the panel is covered with a *resist*—a material that protects selected areas from electrolytic copper plating—except where copper will ultimately remain on the base material. The most common resist material is dry film. *Dry-film plating resist* (See Figure 4-4) is an ultraviolet light-sensitive photo-polymer also called *photoresist*. It's supplied on a roll and applied by processing the panel through heated rollers on a hot-roll laminator.

After hot-roll lamination comes the *exposure step*, in which the panel is placed in a UV printer frame, and a photo tool—either film, glass, or LDI—is positioned to the panel with tooling pins used as locators. The emulsion on the phototool forms the circuitry pattern plus any auxiliary features added during panelization. This emulsion blocks UV light, so the light passes only through the clear portions of the photo tool to activate the light-sensitive resist.

The areas of the dry film exposed to UV light undergo a chemical reaction called *polymerization* and become impervious to the chemical solutions used in the developing process.

The panel is placed in a conveyorized developing machine, which sprays an alkaline solution on the panel to remove the unexposed resist, forming an image of the desired circuitry open to the base copper.

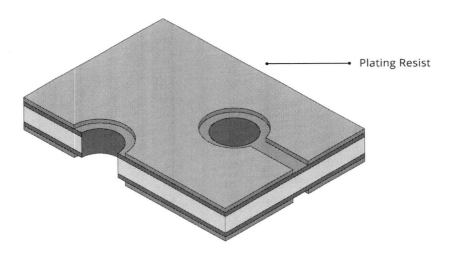

Plating Resist

Figure 4-4: Plating resist is applied, leaving the desired circuitry uncovered.

Step 7: Pattern Plating

In the pattern-plating step, the panels are clamped in plating racks and immersed in a series of chemical baths that clean the copper pattern that makes up the circuitry.

Next, the panels are immersed in a copper-plating solution. The solution and panels have opposite electrolytical charges. These opposite polarities cause copper ions to migrate to the uncoated copper areas on the panel, depositing the desired thickness of copper on the plate's surface and in its holes (see Figure 4-5). This process takes about an hour.

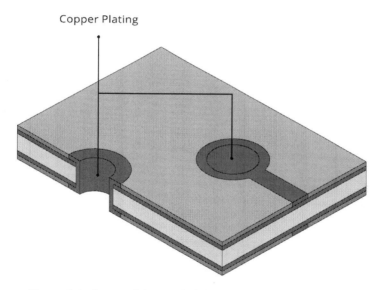

Copper Plating

Figure 4-5: Copper of the specified thickness (usually, 0.0015 of an inch) is deposited electrolytically on the walls of the drilled holes.

After copper plating, the panels are moved from bath to bath, either by hand or by machine. Automatic plating equipment is computer-controlled, and at least one hoist moves the racked panels through the bath sequence without manual intervention. The circuitry pattern, now covered with extra copper, is further electroplated with tin or tin/lead solder (see Figure 4-6).

Etch Resist
(Tin or Tin/Lead)

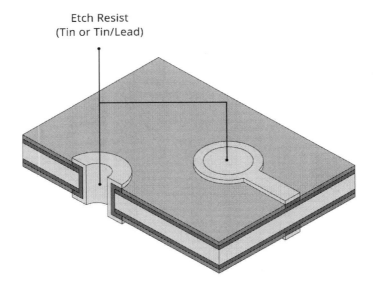

Figure 4-6: Tin or tin/lead is electrolytically deposited over the copper plating.

The tin or tin/lead plating sequence is similar to the copper plating process but requires only a few minutes. Because lead has become environmentally undesirable, the use of tin alone is becoming common.

Finally, the panels are removed from the plating racks. The tin or tin/lead covering the pattern will protect the circuitry in subsequent processing steps.

Step 8: Resist Stripping, Etching, and Chemical Stripping

Next, batches of panels are placed in a tank or processed on conveyorized spray equipment to remove the imaging material. This step is called *resist stripping* (see Figure 4-7).

Exposed Copper Surface

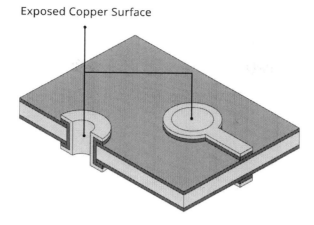

Figure 4-7: Plating resist is chemically removed, revealing the surface copper.

After the resist is stripped off, the panels are placed in a conveyorized spray etcher or batch tank, where a chemical etchant (usually, an ammonia-based compound) removes the uncovered copper but doesn't attack the tin or tin/lead plating, which protects the copper underneath. In this application, the tin or tin/lead plating is called the *etch resist*. (See Figure 4-8).

Then the tin or tin/lead is chemically stripped from the copper, revealing the copper circuitry pattern.

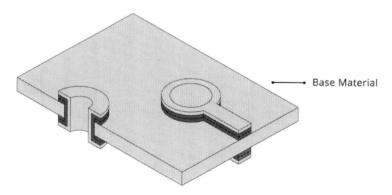

Base Material

Figure 4-8: The unwanted copper is removed chemically by an etchant that attacks copper but not tin or tin/lead plating.

Step 9: Solder Masking

A solder mask (more properly called solder resist) is a nonconductive coating that is selectively processed onto the surface of a PCB to protect certain areas during subsequent operations (see Figure 4-9).

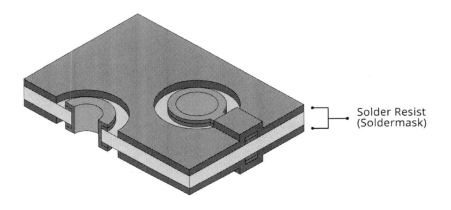

Solder Resist
(Soldermask)

Figure 4-9: The specified solder mask (liquid-photoimageable solder mask or screen-printed solder mask) is applied to the surfaces of the panel.

The areas left uncoated are lands and holes to which components are to be soldered or attached, test points with which electrical probes will make contact, and plug-in contact areas (see "Step 11: Gold and Nickel Plating," later in this chapter).

Liquid-photoimageable solder mask is processed by exposure to UV light through a film or glass phototool. Then the unpolymerized material is removed during developing, as in the dry-film imaging process described in Step 6.

Step 10: Surface Finishing

The fabricator generally supplies the customer with a metallic finish of choice that best suits the needs of the project and to facilitate assembly. Hot-air solder level (HASL) is being used in Figure 4-10.

Hot-Air Solder Level Applied

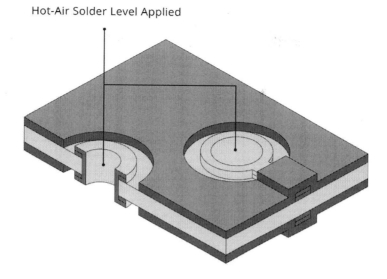

*Figure 4-10: Solder (tin/lead) is applied to the exposed
copper, and the excess solder is removed.*

Step 11: Gold and Nickel Plating

Some PCBs are designed with contact areas, called *fingers* or *tabs,* that ultimately mate with a connecting device. Usually, designers specify nickel and
gold plating for these contact areas.

The first step in the gold plating process is masking all parts of the panel except
the area to be plated, using special adhesive tape that conforms to the circuit
traces and protects them from plating solutions. (Alternatively, the panel is
gold-plated before hot air solder leveling, and the gold fingers are protected
during the hot air solder level process by means of special tape.)

Generally, nickel plating is 100-200 micro-inches thick, and gold plating is
30-50 micro-inches thick. Compared with gold, the nickel surface is harder
and provides better resistance to wear during plugging and unplugging of the
PCB assembly. It also acts as an intermetallic barrier layer that prevents the
gold from migrating into the copper below.

The gold plating process is also known as *tab plating* or *gold tipping*. When unplugging or plugging isn't an issue, numerous PCB designs have replaced the costlier and more labor-intensive hard gold-finger process and switched to a process where the entire PCB is built with the immersion gold finish.

Step 12: Applying the Component Legend

The *component legend*—also called *nomenclature* or *white marking*—comprises the identification symbols that are applied to the board to aid the assembly operation. Test or field service personnel may also use the legend to locate a particular component on an assembled board.

Component legends can be screen-printed on the panel or on individual boards. Designers most often specify epoxy ink, which comes in a variety of colors. The ink is applied with screen-printing technology; then the printed panel or board is cured in an oven or under UV light.

Step 13: Fabrication

The term *fabricate* describes the many mechanical operations that bring the PCB to its final dimensions and create any specified slots, grooves, bevels, or chamfers (see Glossary). The process includes routing based on the Gerber data (see Chapter 3).

One fabrication step involves separating the boards through *scoring*. In this process, V-shaped grooves are machined on opposite sides of a panel to a depth that permits individual boards to be separated from the panel.

Step 14: Electrical Testing

Electrical testing (or e-testing) is the process by which an electrical charge is applied to the circuitry to verify that it functions properly. The test is typically conducted using a test fixture consisting of a frame and a holder containing a field of pins that make electrical contact with the PCB. Dedicated test fixtures are used mainly for production orders, while the flying probe test machines are used for prototype and small production runs. As board complexity continues to increase, the capabilities of dedicated test fixtures are challenged, if not exceeded. Flying probe test machines require no test fixturing, allowing engineers to complete test programs of higher complexity PCBs in less time.

Creating a Test Fixture

The test fixture is designed based on the Gerber-generated test data (see Chapter 3). Usually, it's constructed by drilling a piece of nonconductive material (such as acrylic or epoxy glass) with the same pattern as the drilled holes in the PCB to be tested. For a board that uses surface-mount technology (see SMT discussion in Chapter 1), the pattern of the lands on which the components will be mounted is also drilled.

A test fixture is unique to the PCB that it's designed to test and can't be used for any other board.

Next, spring-loaded metal pins are inserted into the drilled holes of the test fixture to make contact with the corresponding points on the PCB to be tested. The opposite ends of the metal pins are connected by a large pad area to the test equipment. This construction forms a so-called *bed of nails* on which the PCB to be tested is placed.

Putting the Board to the Test

Electrical testers have a top fixture and a bottom fixture to contact both sides of the board. The fixtures are mounted within the electrical tester, and the board or panel is placed on the pins. The Gerber-generated test program causes the test equipment to pass electrical current between specific pairs of pins.

Each pair makes contact with the starting and ending points of a network on the PCB. If no current passes, an open circuit is present; if current passes to an incorrect pin, a short circuit has been detected. Then the test equipment identifies the networks that have shorts and opens.

Today's programs and equipment can test at low or high voltage and can test *insulation resistance*—a measure of the PCB's capability to withstand voltage applied to circuitry on the surface without breaking down (*arcing*).

Flying Probe Testing

A *flying probe* electrical tester tests boards by directing multiple probes to various conductors on the board and sending current from point to point to check for shorts and opens. These devices have numerous probes and are generally slower than dedicated pin testers, which can test all electrical contacts on a board simultaneously.

Flying probes are known for their fast and inexpensive setup and operation costs, which makes them well-suited to quick-turn proto-type (fast turnaround) and golden-board proofing.

Step 15: Final Inspection

The final step in the manufacturing process is a visual inspection of the finished PCBs. Most often, the boards are checked for correct mechanical dimensions, and for overall appearance and other cosmetic parameters. Most fabricators have adopted high operational and quality standards, so final inspection today is more a final audit than a full-blown inspection.

Now that you've seen how double-sided PCBs are made, you're ready to move on to Chapter 5, which discusses the process of manufacturing multi-layer boards.

Want to Save Some Money?

Remember that Standard FR-4 *is* lead-free.

Tg (glass transition temperature) is the temperature at which the resin system changes from a rigid or hard material to a soft or rubber-like material. Some people think only high temperature material can be used in lead-free assembly. But that's not the case.

High Tg material *is* required for the multiple lead-free assembly cycles of higher layer count PCBs, but the standard FR-4 material with lower Tg is more than adequate for the manufacture of single, double, and four-layer PCBs.

Consult a process engineer familiar with your company's PCB assembly process before making any material changes. You could save money by asking whether the more expensive and harder-to-manufacture high Tg material is actually needed for lower layer count PCBs.

Chapter 5

Manufacturing Multilayer Boards

WHILE DOUBLE-SIDED BOARDS once accounted for most of the world's PCB production, multilayer boards (MLBs) are now predominant.

That's because MLBs contain several layers of laminated circuitry that can offer more potential interconnections per unit of area for packaging electronic components. The drive for ever smaller, faster, and more powerful products has pushed these more sophisticated circuit boards—with their additional alignment of artwork (layers)—to the forefront.

The following processes describe the methods used to produce most multilayer PCBs today. Each step can be accomplished in a variety of ways, however, with equipment manufacturers and material suppliers creating their own methods.

Multilayer Fabrication

This section describes the steps used to produce a six-layer board with plated through-holes, solder mask over copper (SMOBC), and solder coating, as well as with gold-plated contacts and component legend. The illustrations depict the changes in the laminate and the surface of the PCB as a multilayer board is processed from material preparation through the solder-coat operation.

To fabricate a six-layer board, as shown in Figure 5-1, two panels of thin, double-sided laminate are imaged and etched to form Layers 2,3,4, and 5. Layers 1 and 6 (the outer layers) are formed with copper foil. This technique is call *foil lamination.*

When the inner layers have been made and sandwiched inside the laminate package, the processing of an MLB is similar, if not identical, to the processing of a double-sided board. Where applicable, we discuss the similarities and differences in the procedures for making both types of boards.

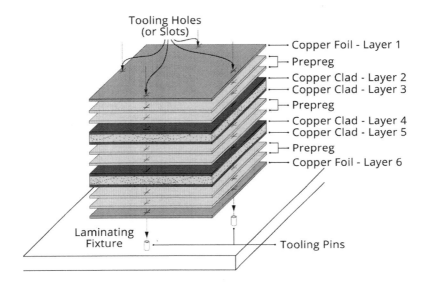

Figure 5-1: Exploded view of a six-layer multilayer board.

Step 1: Preparing Materials

The shop traveler designates the materials to be used for the MLB, including the following information:

- ✓ **Copper:** The traveler defines its thickness and weight.

- ✓ **Epoxy glass:** Different glass weaves and types of epoxy resin are available and are defined in detail in industry specifications (see Chapter 2).

✓ **Panel size:** The thin copper-clad laminate (or core material) used for the inner layers is sheared to panel size (see Figure 5-2), just as would be done for double-sided laminate.

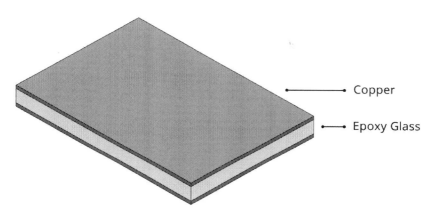

Copper

Epoxy Glass

Figure 5-2: Core material. The copper thickness, epoxy type, and panel size are defined in the shop traveler.

Step 2: Cleaning

The inner layer panels are cleaned—chemically, mechanically, or both—to remove any contaminants from the copper surface. Reverse-current cleaning, or *deplating*, is gaining popularity. The process is somewhat like electroplating, but the polarities are reversed so that copper ions are drawn from, rather than attracted to, the copper.

Step 3: Inner Layer Imaging

Most inner layers of MLBs are imagined with resist applied to the panels (see Figure 5-3). Usually, the resist is dry film, as discussed in Chapter 4. Most often, the layers are printed with the imaging material covering the desired copper circuitry rather than the unwanted copper, which will be etched away.

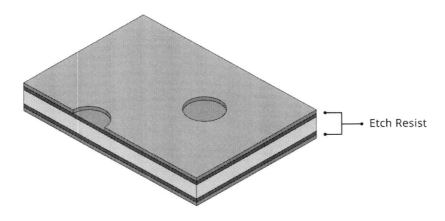

Etch Resist

Figure 5-3: Unwanted copper is exposed.

Step 4: Etch Stripping

Next, the inner layers are chemically etched to remove the unwanted copper. Then the resist is stripped, revealing the copper circuitry.

Step 5: Inspection

As conductor widths have narrowed over time, the visual detection of shorts or opens in inner layers has become more difficult, but also more important. After an inner layer is laminated inside an MLB, the presence of an open or short results in a scrapped board, because repair is almost impossible.

Automated optical inspection (AOI) equipment is capable of finding and identifying flaws that can't be detected by the human eye. Electronic data representing the circuitry can be generated and then downloaded to AOI machines.

AOI equipment also has measuring capabilities, which makes it ideal for use in statistical quality and process control programs.

Step 6: Oxide Treatment

The copper circuitry on an inner layer must be treated before lamination to improve adhesion to the epoxy-glass bonding agents. Improved adhesion also improves structural strength and overall board reliability.

The most common treatment is black or brown oxide plating (see Figure 5-4). Red oxide can also be used. Another option is *double treat*—chemically treated copper foil that improves adhesion much the same way the oxides do.

Oxide-Treated Copper

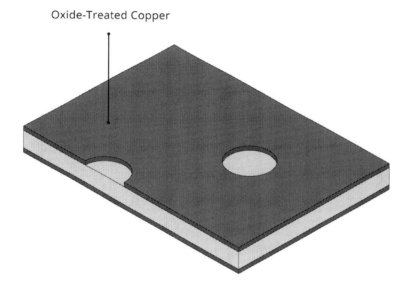

Figure 5-4: The circuitry is treated with an oxide coating.

Step 7: Layup

Refer back to Figure 5-1 to see the stacks that comprise a six-layer MLB package. Layers 2,3,4, and 5 are formed, as discussed in Steps 1 through 6, and the foils that form Layers 1 and 6 are placed on the top and bottom of the stack of materials. The process of creating the stack is called *layup*.

The layups are stacked as much as eight to ten units high and then transferred to a vacuum-laminating press.

Step 8: Lamination

Vacuum lamination, in which the stacks are pressed and heated in a vacuum chamber, reduces the laminating pressure needed and, subsequently, layer-to-layer slippage. The fabricator may use either of two methods:

- ✓ **Vacuum frames:** The stacks are enclosed in vacuum frames built to fit between the platens of a regular press.

- ✓ **Vacuum presses:** The entire press assembly is enclosed, and a vacuum chamber is formed.

Fabricators use sheets of aluminum or stainless steel plus stick-resistant plastic between individual layups to protect the copper foil and keep the layups from sticking together.

The prepreg—partially cured epoxy glass—provides the bonding material between layers. Because the prepreg is only partially cured, heat and pressure cause it to flow and bond to the surface of the inner layers during the lamination operation.

The laminating press first applies a vacuum, removing all the entrapped air and gases, and then applies heat and pressure to the stack so that the thermosetting resin in the prepreg undergoes a molecular bonding known as cross-linking that binds the layers tight (see Figure 5-5).

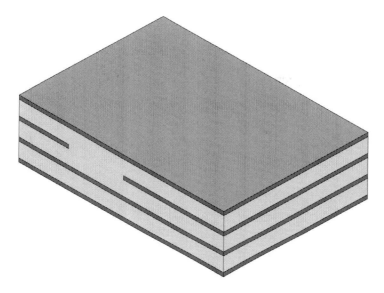

Figure 5-5: Heat and pressure cause prepreg to flow and bond the layers.

Following the heating and pressing cycle (which usually lasts about an hour, depending on the equipment used), the panels are transferred to a cooling station, where they're clamped under pressure until they cool.

Step 9: Stress Relief

Next, the cooled panels are placed on baking racks and put in an oven. The high heat and pressure to which the panels have been subjected create internal stresses that, if not relieved, will cause them to warp and twist. Oven heat, usually at 325 degrees Fahrenheit for two to four hours, relieves these stresses.

Step 10: Fabricating Tooling Holes and Trim Edges

Next, the edges of the panel are trimmed with a router (manual or automatic), a diamond saw and template, a milling machine, or some other mechanical process. The trimming operation eliminates the possibility of any prepreg *flash* (outflow) and smooths any uneven or rough edges.

After trimming, tooling holes are created (see Figure 5-6). The operator may use any of the following methods to create tooling holes:

- **Pin lamination:** In some operations, a panel is fabricated with tooling pins in the layup during the lamination cycle. The panel has a series of holes along its edges. In this situation, the holes are used as a reference to locate the panels on the drill machine table. The holes are also used as a reference for trimming the panels to the size specified in the traveler.

- **X-ray:** In other cases, the tooling holes are formed with X-ray techniques. Lands or targets on the inner layer are located by X-ray, and then the corresponding tooling holes are punched or drilled.

- **Bombsight drilling**: Another method of forming the tooling holes is *bombsight drilling* to internal target patterns. The patterns are exposed by selectively machining away the copper foil to expose a target on the inner layer.

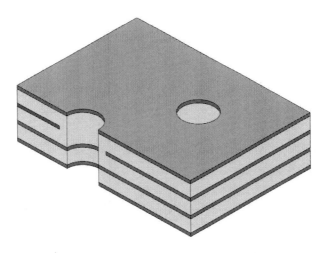

Figure 5-6: Hole sizes and locations are determined by drill data furnished by the customer.

Step 11: Drilling

Drilling MLBs is essentially identical to drilling double-sided boards, except that MLBs must have smooth, smear-free holes. As the drill bit passes through layers of copper and epoxy glass, it generates heat, which can cause the epoxy to smear along the sides of the hole. The subsequent copper plating must form a reliable bond with the internal copper circuitry, so the holes must be free of drilling dust, debris, and smear. (See Step 13 for more information on the removal of drilling smear.)

Higher-density MLB designs often have smaller tooling holes than double-sided designs do. To maintain quality, fabricators work with smaller stacks. Because MLBs have buried circuitry, many fabricators use X-ray equipment to make sure that a drilled hole is aligned with an internal land. X-ray machines can also locate multiple lands on the panel and change the drilling program to align with the internal circuitry.

Several types of drills are used for making *vias* (small holes, discussed in Chapter 1) in MLBs. The most common type remains the mechanical drill, which uses a series of spindles, each with bits that operate in the Z axis (up and down) to form holes. Smaller holes known as *microvias* are typically formed with a laser drill (see Figure 5-7). The main types of laser drills in use today include infrared (also referred to as CO_2) and ultraviolet (UV). Some laser drilling systems feature a combination of the two types.

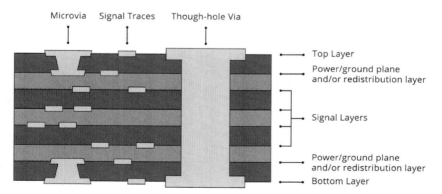

Figure 5-7: Operators use laser drills to create microvias.

Step 12: Deburring

Drill burrs are removed from MLBs the same way they are from double-sided boards, but additional care is required because of the smaller holes in MLBs, which are more likely to retain drilling debris. If holes are clogged, plating solution can't flow through them and deposit copper. Partial or no copper deposition in holes means little or no electrical connection to the circuitry on the inner layers, which would cause the boards to be rejected. A high-pressure washer is used to remove any debris from the holes.

Step 13: Electroless Copper Plating

An MLB requires several additional process steps to complete copper deposition because the drilling process may have created epoxy smear. This smear prevents the electroless copper from bonding to the copper on the inner layers, thus creating an open circuit or an unreliable plated through-hole.

The epoxy smear is removed by immersing the panel in a series of chemical solutions, followed by potassium permanganate or concentrated sulfuric acid. Smear can also be removed by means of *plasma*—a dry chemical method in which the panels are exposed to oxygen and fluorocarbon gases.

After desmearing, the panels are processed in the same way as double-sided products, with copper plating being applied (see Figure 5-8).

Electrodeposited Copper

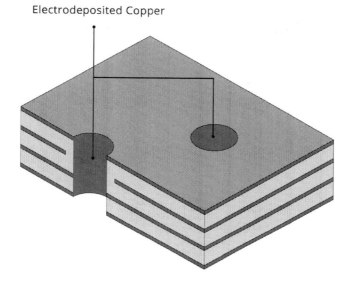

*Figure 5-8: A thin layer of copper is deposited follow-
ing smear removal, cleaning, and preparation.*

Steps 14 through 24

These steps in MLB manufacturing include:

- Imaging

- pattern plating

- resist stripping, etching, and chemical stripping

- inspection

- solder masking

- solder coating

- gold plating

- component legend

- final fabrication

- electrical testing

- final inspection

Other than the additional care required to create a more expensive product, the remaining operations are the same for MLBs as for double-sided boards, as described in Chapter 4.

High-Tech Aspects of Multilayer Boards

Much of the process for the latter part of creating an MLB is similar to creating a double-sided board, but a few things are distinct. We go over some of the differences in the following sections.

Better Plating Through Chemistry

As layer counts increase and hole sizes decrease, the *aspect ratio* (the ratio of the panel thickness to the hole diameter) increases dramatically. This higher ratio demands special techniques and chemistries to achieve uniform plating and adequate thickness in the hole. New agitation systems and solution-movement methods have been created to increase the flow of solution through the holes, thus improving plating uniformity. Also, new chemistries have been developed to enhance the uniformity and quality of plating deposition in smaller holes.

A plating method known as *pulse plating* uses pulses rather than direct current to plate the copper. Changing the pulse width and height allows fabricators to plate at a faster rate than they can by using direct current. Some pulse-plating units feature a short reverse pulse that cleans the copper surface and permits very smooth plating.

Controlled Impedance

More and more of today's PCB designs demand faster information processing. Yet getting a signal from one place to another, even in such a tiny device as a PCB, can come up against resistance. This resistance is known as *impedance*.

Think of impedance as pot holes or speed bumps in the road that a signal needs to travel through on the PCB.

With that in mind, many customers these days are requesting *controlled imped-ance*—the elimination or minimizing of discontinuities in the signal that a trace (conductor) delivers. With controlled impedance, transmission lines are designed to transport energy at high speeds with little loss in signal shape, mag-nitude, or speed to connect a plated through-hole to a via or to another device.

Be aware that controlled impedance adds another level of complexity to the design, artwork, material selection, and PCB manufacturing process. Upon receiving an order with a controlled impedance request, the vendor will con-duct a simulation to verify that the design will allow for the final impedance values (with a usual tolerance of plus or minus 10 percent).

If the design calculations don't meet the standard required, the PCB manu-facturer will notify the customer and request a change in the stackup to meet the designed needs. Artwork being plotted has to compensate to allow for the plating tolerances of the PCB manufacturer. Material consistency is also important—glass style, resin content, and resin flow affect the dielectric con-stant, which will affect the impedance results.

All controlled impedance boards require TDR (Time Domain Reflectometry) measurements, which confirm that the impedance values are within tolerance. To make this measurement, TDR traces are placed on a coupon located on the manufacturing panel. The coupon usually contains several traces (usually up to 8 inches long) that run parallel to each other, with a plated through-hole at either end. Using a TDR measuring device, the coupon verifies that the impedance values are met.

The testing is done on the coupon and is considered the benchmark that both designer and fabricator agree upon, because the traces in the PCB are gener-ally not easily accessible or are too short for testing. You can ask for the TDR report, along with the coupons, to accompany the shipment.

Blind and Buried Vias

One significant processing sequence variance involves design techniques called *blind and buried via holes* (see Figure 5-9).

- **Blind Via:** The plated hole doesn't extend through the entire thickness of the board but stops at a given layer.

- **Buried Via:** A plated through-hole located within an inner layer doesn't extend to either external surface.

Manufacturing MLBs that call for buried vias isn't considered to be difficult. An inner layer can be processed as a double-sided board with plated through-holes and then pressed in a multilayer, after which the internal vias would be buried.

Blind vias, however, require the operator to drill to a given depth in the panel, but no deeper—a technique known as *controlled-depth drilling.* Special drilling equipment is required, and the process is more expensive than conventional drilling because the panels can't be stacked. Also, specific plating operations are required to ensure correct plating to the bottom of a blind hole. Still, as semiconductor packaging demands more interconnections, the number of designs using blind and buried via holes is increasing.

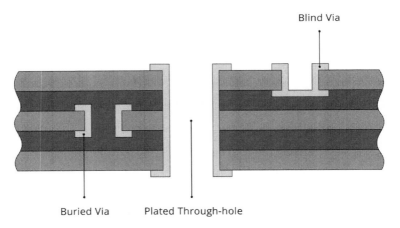

Figure 5-9: Some operations require buried via holes (left) instead of the blind kind.

High-Density Interconnection (HDI) Technology

The use of more complex components essential for mobile communications and computer chip packaging has pushed PCB fabricators to produce high-density interconnection (HDI) boards. HDI boards have a higher circuitry density than traditional circuit boards, contain blind and/or buried vias, and often contain microvias that give PCB designers the option of placing more components on either side of the bare board. These products are typically small, light, and thin.

As shown in Figure 5-10, HDI boards require additional manufacturing technologies like sequential lamination, where an already laminated sub-part known as "n" is laminated to an additional layer or layers of copper.

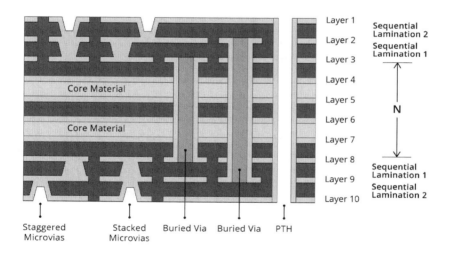

Figure 5-10: Cross-section of an HDI board showing micro-vias, blind vias and sequential lamination.

Using industry terms, a 1-n-1 PCB contains a single buildup of a high-density interconnected layer (the "n") that requires one sequential lamination on each side of the core. The 2-n-2 PCB has two additional HDI layers sequentially laminated, and usually incorporates copper-filled stacked or staggered micro-via structures.

Not every board house can build HDI circuit boards. Careful research and vetting procedures are required before placing an HDI order with your PCB supplier. Successful manufacturing of HDI PCBs requires special equipment and processes such as laser drills, via plugging, laser direct imaging and sequential lamination cycles. HDI boards have thinner lines, tighter spacing and tighter annular rings, and use thinner specialty materials.

Successfully producing this type of board requires additional time and a significant investment in manufacturing processes and equipment.

Keeping PCBs Dry

IPC-1601: Printed Board Handling and Storage Guidelines provides suggestions for proper handling, packaging, and storage of PCBs. This document gives the PCB supplier total responsibility for PCB moisture content even after the finished product has left the manufacturing facility.

A bare board begins to absorb moisture immediately upon leaving the factory. The amount of moisture absorbed depends on a variety of factors, including:

✓ Base material used in manufacturing.

✓ Manufacturing environment.

✓ Packaging method.

✓ Shipping temperatures (from the cold bellies of aircraft to hot delivery trucks).

✓ Customer storage and inventory procedures.

Vacuum sealing and the use of desiccant only delay or lessen moisture absorption; they don't prevent it. The longer a PCB is stored on a customer's shelf, the greater the chance it will absorb moisture, which can manifest itself in the assembly operation as *delamination*.

Delamination is caused either by moisture or by manufacturing defects. If a problem PCB is determined to be structurally sound, the problem most likely is moisture-related, and a bake-out process before any additional assembly can remove most of the moisture, if not all of it.

As the demand for lead-free materials continues to increase, it's highly advisable to bake the bare boards before any assembly operation, especially if the PCBs have been in storage for more than six months.

Chapter 6

Should I Buy My Boards Offshore?

GETTING PCBS OFFSHORE is now routine in the circuit board business because of significant savings. But there are also some potential risks and roadblocks you should be aware of if you want to avoid serious repercussions.

So how do you decide where to buy your boards?

First, take a critical look at your business. Determine which jobs make up the bulk of your product and which make up the majority of your profit. Then ask these questions:

- **Is it a board related to the military?**
  Because of the International Traffic in Arms Regulation (ITAR), PCBs specifically designed to meet military specifications or modified for military use must stay domestic (in the United States) for manufacture. There are plenty of domestic manufacturers qualified to provide quality PCBs that meet military specifications.

Which way should you go?

- **Are forecasts available?** Forecasting provides an enhanced opportunity to meet customer demand and may even give you better pricing opportunities. If your customer cannot give you accurate forecasts, offshore may not be the way to go. Alternatively, you may want to consider *both* domestic and offshore production to meet customer spikes in demand. Realize that some jobs will always need to be produced onshore, while others are better suited to run with an offshore partner.

- **What kind of volume are you looking at?** The number of boards required and the timeframe you're working within may affect your decision.

- **Prototype:** This would be several pieces in a matter of days. Generally, if you need it in one to four days, you want to keep these orders domestic. But if you can wait five to seven days, prototypes can easily go offshore for better pricing.

- **Low-to-medium volume:** Here, we're talking about anywhere from approximately 50 pieces to several hundred pieces at a time in two to four weeks, *or* several thousand pieces over a year's time. You should be able to find a domestic manufacturer to build this order size within the required timeframe and with the quality desired. Depending on your customer requirements, however, offshore manufacturing offers a compelling and usually more cost-effective alternative.

- **High volume:** This would be hundreds or thousands of pieces at a time, produced on a definite schedule. Intended for automotive or a wide range of consumer products, these PCB orders are prime candidates for offshore manufacture.

- **Does the board have high quality requirements?** The tougher the quality requirements, the longer the learning curve for any potential supplier, especially when it comes to aerospace and medical applications. Not every manufacturer can build to IPC-6012 Class 3. Keep in mind that if you go offshore, you will have to clearly communicate all manufacturing needs, including the particular nuances of each board's requirements, to a supplier whose native language may not be English.

Buying Offshore

Whether you are a purchasing manager for an original equipment manufacturer, a buyer for a contract manufacturer, or the owner of a domestic North American PCB factory, partnering with an offshore supplier—either directly or indirectly—can help you meet customers' expanding demands.

Even when our industry is dealing with trade war tensions and other obstacles, buying boards overseas still makes financial sense. By doing so, you can save a lot of money.

Before you take the plunge, however, carefully consider whether your company is a big enough fish to swim safely in the offshore manufacturing ocean. The more orders you can place with an overseas vendor, the more weight you'll have to throw around when necessary.

If, after careful consideration, you decide that offshore manufacturing makes sense for a particular PCB, there are a couple of established methods to procure offshore PCBs. Each has its advantages and disadvantages. The crucial factor in deciding which method to use is whether you intend to use overseas vendors for only some or for most of your PCB requirements.

Need help navigating the offshore waters? Better Board Buying (B3) makes it easy. Let us help you.

Option 1: Use a PCB Broker

To minimize the challenges of getting quality circuit boards delivered on time and at the right price, many companies work with PCB distributors, also known as brokers.

PCB distributor or supplier firms are usually staffed by sales people who have offices (distributorships) in the United States, as well as near the offshore factories. Brokers eliminate much of the pain involved in doing business with offshore vendors, assuming full responsibility for quality, responsiveness, and on-time delivery. Some even manage inventory.

Working with a distributor can save you time and money in the following ways:

- **You don't have to qualify vendors.** The distributor can prequalify potential offshore partners, screening for proper credentials (such as UL, ISO, and TS certification) and checking references. Its reps can also go to the factories to survey and inspect the facilities in person.

- **The distributor can determine which part numbers would be better produced offshore.** Not all part numbers are alike, and offshore production may not be the best solution for every part number. The distributor can identify which products are best produced overseas and may even warehouse part numbers to help you manage your inventory costs.

- **You don't have to fill out the qualification order.** Many engineering (and other) questions must be asked and answered to ensure the production of a high-quality board. The distributor can take care of all the details necessary to get your orders delivered on time, in good shape, and with all the required paperwork.

- **You gain economies of scale.** The distribution firm often has more buying power and—even more important in today's world of rising freight costs—greater shipping power. The total cost of acquisition through a distributor may be lower than you think.

- **You don't have to work directly with the vendors.** You don't have to stay up until all hours of the night trying to communicate with vendors in time zones 12 hours ahead of yours. The distributor's reps are there to deal with the vendors, handling quotes, and solving engineering problems and scheduling challenges. They can answer your questions as well.

- **You don't have to worry about quality concerns or delivery problems.** The broker is the ideal go-between to handle any quality or delivery problems with an offshore manufacturer.

- **You save on bank fees.** The distributor will grant you terms and take payment by check or some form of electronic payment, while overseas vendors often must be paid in advance via expensive international wire transfers.

Want more information about using a PCB broker? Visit BoardBuying.com.

Option 2: Go Direct - Establish a Relationship with an Overseas Supplier

One offshore option that may give your company the best pricing is to work directly with an overseas PCB manufacturer. Cutting out the middleman in your supply chain can reduce your costs and give you a leg up on your competition.

Going direct will, however, require time and effort on your part, including frequent visits by your senior personnel to the host country to establish and maintain strong relationships with your vendors. A good inside contact at the supplier—one who thoroughly understands the requirements of your operation—will be crucial.

Also, consider whether this partner will be able to single-handedly meet your future needs. You may have to establish a relationship with more than one offshore supplier.

Offshore PCB manufacturing has matured a great deal in the last decade. It is not all that difficult anymore to deal directly with overseas vendors. It does, however, sometimes come with unexpected administrative costs and headaches, such as freight bills, and dealing with quality problems and customs issues. With a bit more work on your part, though, you can expect significant savings, with a lower total cost of ownership for your PCB orders.

Want more information about going direct? B3 can help!

Chapter 7

Successful PCB Partnerships

NO MATTER WHICH sourcing option you choose for your PCB supply chain, it's crucial that you follow best practices to ensure a thriving partnership. These tips will help you make good choices and build successful business relationships in the printed circuit board industry, both before you take on a supplier and/or distributor, and during the course of your alliance.

1. **Check the Supplier's Financial Fitness:**
    Ensure that your PCB supplier is financially sound. Pull a credit history and possibly even talk to the supplier's bank. A vendor cannot deliver product in three to four weeks if it can't purchase the materials required to build a customer order the day it is placed.

    Obviously, this will be easier to do with U.S.-based suppliers. For overseas manufacturers, if you cannot investigate the vendor's financial fitness, double up on reference checks.

2. **Follow Up on References:**
    It's common sense, but surprisingly, many PCB buyers neglect this crucial step. Don't be one of those buyers. Checking references is vital when considering a new PCB supplier. You can learn a lot by asking a vendor's references detailed questions about product quality

and delivery. And when you ask a vendor for references, confirm that the supplied references are using the same technology that you will need for your orders. Some suppliers may have great references for lower-tech projects but may not be up to par for those that require a higher technology level. You don't want to find that out after you've already placed an order.

3. **Spread Out PCB Requirements:**
   Not all PCB manufacturers are equal. Don't push your supplier's technology envelope. Let each supplier fulfill orders with the technology that it can build comfortably. Adding another PCB supplier to your vendor base is a lot less work than overcoming an ongoing quality and/or delivery situation. And not all PCB manufacturers can build everything you need. Don't try to force a production supplier to build prototypes, or a prototype supplier to build production volumes. Give the vendor the business that makes sense for its specific operation.

4. **Communication Early and Often:**
   This might sound obvious, but PCB buyers often do not communicate with vendors until there is a problem. Both the PCB supplier and the board customer should feel comfortable communicating at all stages of an order. Feedback is a gift. Let your PCB suppliers know when their performance is not acceptable and, just as importantly, when they are doing well. And encourage them to be straightforward with you throughout an order's life cycle.

5. **Get to Know Your Vendor:**
   Although customer service is the name of the game for PCB vendors, it does take two to tango. Suppliers are just like everyone else; they prefer to work with those they know personally and have had good experiences with. They'll often work harder for customers they like, even when it's not expected or required. Invest some time in getting

to know your suppliers (and not just one person at a particular company). It will greatly benefit your supply chain.

6. **Develop a Problem Resolution Program:**
   PCB buyers often invest a lot of time at the beginning of the PCB supplier relationship, from signing a non-disclosure agreement (NDA), to spelling out manufacturing specifications, to sending quotes. But too often, they don't put enough time into developing a procedure with each PCB vendor for handling the inevitable quality problems.

   The PCB is a custom-made item that involves over 100 individual processes performed by both humans and machines. In this industry, it is not *if* but *when* a quality concern will occur. Having a documented plan of action in place when a quality problem does occur with an order—one that is clearly understood by both parties—will help resolve quality concerns much faster and more easily. Set one up at the start of your relationship and make sure the PCB supplier is fully onboard with the idea.

7. **Be Wary of Self-Inflicted Costs:**
   Be smart about how you order PCBs when it comes to delivery, quantity, technology, and quality inspection. Talk to as many people from your company as you can and try to find out what is truly needed, and whether the required delivery date is realistic. The more you know, the lower your costs and the better your planning and buying performance.

   You should be asking questions like these:

   • **Do we really need the PCB in five days when the components won't arrive for another week?** PCB costs rise exponentially when the delivery window shrinks from weeks to days.

   • **Should we order the same quantity of the same board every month or should we order quarterly or even yearly?** There is definitely a cost savings when more of the same product is manufactured at the same time.

- **Do we design the most high-tech PCB that only a few suppliers can manufacture? Or do we design a PCB that is easier to manufacture?** A Ford will get you there just like a Porsche, but at a fraction of the cost.

- **Does that circuit board really need to be Class 3 IPC-6012 or just IPC-A-600?** There is a tremendous cost difference between these two standards because of additional testing and paperwork required for Class 3.

8. **Stay Professional:**

As in any industry (or area of life), problems will occur. When they do, the best course of action is to deal with each issue calmly and factually. Don't let your negative emotions ruin your company's relationship with the PCB vendor. A positive attitude and display of leadership will solve problems faster and allow you to strengthen the relationship.

9. **Quote for Fun:**

As a buyer, it's your responsibility to keep your vendor base competitive in pricing as your business relationship continues. It's easy to get complacent when a vendor's quality and delivery are satisfactory. But if you're not regularly checking the pricing of your existing vendors against that of potential suppliers, you could be leaving money on the table.

If quality and delivery performance among your present vendors is comparable, you need a way to ensure that you're still getting the best prices. One way to do this is what could be called "quoting for fun."

Don't hesitate to give potential vendors who are seeking your business a shot at quoting an ongoing project "for fun." Let potential vendors know it's an existing project and you'd like to see where they stand on pricing. It will provide a useful benchmark against which to measure your current vendors' pricing.

And if there is a significant per board price difference—especially if you're having any quality or delivery issues with a current vendor—it may turn into more than just "fun." It may give you another vendor to add to your base. While a few pennies per board difference in price is not worth switching vendors if you're happy with existing ones, it's wise to keep your options open. It will also help keep your vendor base on its collective toes.

Let your vendors know right at the beginning of your relationship that you'll be regularly evaluating their performance in all areas, including pricing.

10. **Conduct Regular Surveys:**

As the customer, you should physically inspect your PCB supplier's operation throughout your business relationship. Surveys help improve accountability, communication, and expectations. You can review technologies and capabilities, and often you will discover potential problems that can be avoided. This will strengthen your relationship with your supplier and keep your PCB supply chain humming.

11. **Pay Promptly**

Last, but certainly not least, *pay your PCB supplier on time*. Payment ensures financial stability and means uninterrupted flow of product. Remember, customers are no longer customers when they don't pay. You won't get the same level of service if you stop being a reliable customer.

Better Board Buying specializes in offshore PCB vendor management. We can help you create and manage a robust supplier base.

Glossary

Terms Commonly Used in the PCB Industry

activating: A treatment that renders nonconductive material receptive to electroless deposition.

additive process: A process for obtaining conductive patterns by the selective deposition of conductive material on clad or unclad base material.

annular ring: That portion of conductive material completely surrounding a hole.

array: A group of elements or circuits (or circuit boards) arranged in rows and columns on a base material.

artwork: An accurately scaled configuration used to produce the artwork master or production master.

artwork master: The photographic film or glass plate that embodies the image of the PCB pattern, usually on a 1:1 scale.

aspect ratio: A ratio of the PCB thickness to the diameter of the smallest hole.

assembly: A number of parts, subassemblies, or any combination thereof, joined together.

automated optical inspection (AOI): Visual inspection of the circuit board using a machine scanner to assess workmanship quality.

automatic test equipment (ATE): Equipment that automatically analyzes functional or static parameters to evaluate performance.

backup material: A layer composed of phenolic, paper composite, or aluminum foil-clad fiber composite used during fabrication to prevent burrs and to protect the drill table.

ball grid array (BGA): An IC package with a large number of terminations arranged in a matrix on the bottom of the package. Connections are made through solder terminations on the underside of the array in the form of solder balls.

barrel: The cylinder formed by plating through a drilled hole.

base copper: The thin copper foil portion of a copper-clad laminate for PCBs. It can be present on one or both sides of the board.

base material: The insulating material upon which a conductive pattern may be formed. It may be rigid or flexible or both. It may be a dielectric or insulated metal sheet.

base material thickness: The thickness of the base material excluding metal foil or material deposited on the surface.

bed-of-nails fixture: A test fixture consisting of a frame and a holder containing a field of spring-loaded pins that make electrical contact with a planar test object (for example, a PCB).

bevel: An angled edge of a printed board.

bleeding: A condition in which a plated hole discharges process materials of solutions from voids and crevices.

blind via: A conductive surface hole that connects an outer layer with an inner layer of a multilayer board without penetrating the entire board.

blister: A localized swelling and separation between any of the layers of a laminated base material, or between base material or conductive foil. It is a form of delamination.

bond strength: The force per unit area required to separate two adjacent layers of a board by a force perpendicular to the board surface.

book: A specified number of stacks of prepreg plies (sheets), which are assembled for curing in a lamination press. See also *prepreg*.

bow: The deviation from flatness of a board, characterized by a roughly cylindrical or spherical curvature such that if the board is rectangular, its four corners are in the same plane.

breakout: A condition in which a PCB hole isn't completely surrounded by the land or annular ring.

B-stage material: Sheet material impregnated with a resin cured to an intermediate stage (B-stage resin). *Prepreg* is the popular term.

B-stage resin: A thermosetting resin that is in an intermediate state of cure.

buried via: A via hole that doesn't extend to the surface of a printed board.

burr: A ridge left on the outside copper surface after drilling.

CAD data: Electronic data that describes the placement of holes and connectors in a PCB. See also *computer-aided design*.

CAM: See *computer-aided manufacturing*.

capacitance: The property of a system of conductors and dielectrics that permits storage of electricity when potential differences exist between conductors.

chamfer: A broken corner to eliminate an otherwise sharp edge.

circuit: The interconnection of a number of devices in one or more closed paths to perform a desired electrical or electronic function.

circuitry layer: A layer of a printed board containing conductors, including ground and voltage planes.

clad or cladding: A relatively thin layer or sheet of metal foil that is bonded to a laminate core to form the base material for printed circuits.

cleanroom: A room in which the concentration of airborne particles is controlled to specified limits.

coefficient of thermal expansion (CTE): The measure of the amount a material changes in any axis per degree of temperature change.

component: An electronic device, typically a resistor, capacitor, inductor, or integrated circuit, that is mounted to the circuit board and performs a specific electrical function.

component hole: A hole used for the attachment and electrical connection of a component termination, such as a pin or wire, to the circuit board.

component side: The side of the circuit board on which most of the components will be located. Also called the *top side*.

composite epoxy material (CEM): A punchable material (paper) used in single-sided boards, but not suited for plated through-holes.

computer-aided design (CAD): A software program with algorithms for drafting and modeling, providing a graphical representation of a printed board's conductor layout and signal routes.

computer-aided manufacturing (CAM): The use of computers to analyze and transfer an electronic design (CAD) to the manufacturing floor.

computer-integrated manufacturing (CIM): Software that takes assembly data from a CAD or CAM package and, using a pre-defined factory modeling system, outputs routing of components to machine programming points, and assembly and inspection documentation.

conductor: A thin conductive area on a PCB surface or internal layer usually composed of lands (to which component leads are connected) and paths (traces).

conductor spacing: The distance between adjacent edges (not centerline to centerline) of isolated conductive patterns in a conductor layer.

conductor thickness: The thickness of the conductor, including all metallic coatings.

conformal coating: An insulating protective coating that conforms to the configuration of the object coated and is applied on the completed board assembly.

connector area: The portion of the circuit board that is used for providing electrical connections.

contaminant: An impurity or foreign substance that, when present on an assembly, could electrolytically, chemically, or galvanically corrode the system.

continuity test: A test for the presence of current flow between two or more interconnected points.

controlled impedance: The matching of substrate material properties with trace dimensions and locations to create specific electric impedance as seen by a signal on the trace.

core thickness: The thickness of the laminate base without copper.

C-stage: The condition of a resin polymer when it is in a solid state with high molecular weight.

curing: The act of applying heat to a material to produce a bond.

deburring: Process of removing burrs after drilling.

defect: Any nonconformance to specified requirements by a unit or product.

definition: The fidelity of reproduction of pattern edges, especially in a printed circuit, relative to the original master pattern.

delamination: A separation between any of the layers of the base of laminate, or between the laminate and the metal cladding originating from, or extending to, the edges of a hole or edge of a board.

design for manufacture (manufacturability) (DfM): Designing a product to be produced in the most efficient manner possible in terms of time, cost, and resources; taking into consideration how the product will be processed; and using the existing skill base to achieve the highest yields possible.

design rule: Guidelines that determine automatic conductor routing behavior with respect to specified design parameters.

design rule checking: The use of a computer program to perform continuity verification of all conductor routing in accordance with appropriate design rules.

desmear: The removal of friction-melted resin and drilling debris from a hole wall.

dewetting: A condition that results when molten solder has coated a surface and then receded, leaving irregularly shaped mounds separated by areas covered with a thin solder film and with the base material not exposed.

dielectric: An insulating medium that occupies the region between two conductors.

digitizing: The converting of feature locations on a flat plane to a digital representation in X-Y coordinates.

dimensional stability: A measure of the dimensional change of a material that is caused by factors such as temperature changes, humidity changes, chemical treatment, and stress exposure.

double-sided board: A printed board with a conductive pattern on both sides.

drilling: The act of forming holes (vias) in a substrate by mechanical or laser means.

dry-film resist: Coating material specifically designed for use in the manufacture of printed circuit boards and chemically machined parts. They are suitable for all photomechanical operations and are resistant to various electroplating and etching processes.

dry-film solder mask: Coating material applied to the printed circuit board via a lamination process to protect the board from solder or plating. See also *dry-film resist.*

electrodeposition: See *electroplating.*

electroless copper: A thin layer of copper deposited on the plastic or metallic surface of a PCB from an autocatalytic plating solution (without the application of electrical current).

electroless immersion gold (ENIG): One of the best and most common

RoHS finishing methods for printed circuit boards. ENIG chemistries offer many strengths, including excellent wettability, coplanarity, surface oxidation, and long shelf life.

electroplating: The electrodeposition of an adherent metal coating on a conductive object. The object to be plated is placed in an electrolyte and connected to one terminal of a direct-current voltage source. The metal to be deposited is similarly immersed and connected to the other terminal.

entry material: A thin layer of material composed of phenolic, aluminum foil, or paper that is placed on top of the panel prior to drilling, to improve drill accuracy, and prevent burrs and dents.

epoxy: A family of thermosetting resins. Epoxies form a chemical bond to many metal surfaces.

epoxy smear: Epoxy resin that has been deposited on edges of copper in holes during drilling, either as uniform coating or in scattered patches. It is undesirable because it can electrically isolate the conductive layers from the plated through-hole interconnections.

etchback: The controlled removal of all components of the base material by a chemical process acting on the sidewalls of plated-through holes to expose additional internal conductor areas.

etching: The chemical, or chemical and electrolytic, removal of unwanted portions of conductive materials.

fabrication drawing: A drawing used to aid the construction of a printed board. It shows all of the locations of the holes to be drilled, their sizes and tolerances, dimensions of the board edges, and notes on the materials and methods to be used. It is called "fab drawing" for short. It relates the board edge to at least one hole location as a reference point, so that the NC drill file can be properly lined up.

first article: A sample part or assembly manufactured prior to the start of production for the purpose of ensuring that the manufacturer is capable of delivering a product that will meet specified requirements.

first article assembly report (FAAR): The DirectPCB validation for component reimbursement of the complete assembly to ensure that the assembly process, the components, and the PCB are successful.

fiducial (mark): A geometric feature incorporated into the artwork of a printed wiring board or into stencils that are particularly necessary for the accurate placement of fine-pitch components.

flex circuit: Printed circuitry that utilizes flexible rather than rigid laminate material.

flux: A chemically active agent that speeds the wetting process of metals with molten solder.

flux residue: A by-product of the soldering operation which may or may not need to be removed from the board.

flying probe: A testing device that uses multiple moving pins to make contact with two spots on the electrical circuit and send a signal between them, a procedure that determines whether faults exist.

FR-1: A paper material with a phenolic resin binder. FR-1 has a Tg (glass transition temperature) of about 130 degrees C. See also *Tg*.

FR-2: A paper material with phenolic resin binder similar to FR-1, but with a Tg (glass transition temperature) of about 105 degrees C. See also *Tg*.

FR-3: A paper material that is similar to FR-2, except that an epoxy resin is used instead of phenolic resin as a binder. Used mainly in Europe.

FR-4: The UL-designated rating for a laminate composed of glass and epoxy that meets a specific standard for fire-retardance. FR-4 is the most common dielectric material used in the construction of PCBs.

front end: Work performed on a CAM system. See also *computer-aided manufacturing.*

G10: A laminate consisting of woven epoxy-glass cloth impregnated with epoxy resin under pressure and heat. G10 lacks the anti-flammability properties of FR-4. Used mainly for thin circuits such as in watches.

Gerber: A software format used by the photoplotter to describe the printed circuit board design.

gold finger: A portion of a conductive pattern (usually gold-plated) on or near any edge of a printed board that is intended for mating with an edge board connector.

golden board: A board or assembly that is verified to be free of defects.

ground plane: A conductor layer, or portion of a conductor layer, used as a common reference point for circuit returns, shielding, or heat sinking.

high-density interconnect (HDI): Ultra-fine geometry multilayer PCB constructed with conductive microvia connections. These boards also usually include buried and/or blind vias and are made by sequential lamination.

hole breakout: A condition in which a hole is partially surrounded by the land. See also *lands.*

hole pattern: The arrangement of all holes in a printed board with respect to a reference point.

hot air solder leveling (HASL): A method of coating exposed copper with solder by inserting a panel into a bath of molten solder, then passing the panel rapidly past jets of hot air.

imaging: The process by which panelization data are transferred to the photoplotter, which in turn uses light to transfer a negative image circuitry pattern onto the panel.

immersion silver: A substance deposited directly on the copper surface by a chemical displacement reaction, which provides a RoHS-compliant finish for printed circuit boards. It is gaining popularity in the industry for its excellent wettability and coplanarity properties.

impedance: The total passive opposition offered to the flow of electric current. This term is generally used to describe high-frequency circuit boards.

in-circuit test (ICT): An electrical test of an assembly in which each component is tested individually for electrical function, to verify correct placement and orientation.

inkjetting: The dispersal of well-defined ink "dots" onto a PCB. Inkjet equipment uses heat to liquefy a solid ink pellet and change the ink into a liquid, which is then dropped via a nozzle onto the printed surface, where it quickly dries.

inner layers: The internal layers of laminate and metal foil within a multilayer board.

insulation resistance: The electrical resistance of an insulating material that is determined under specific conditions between any pair of contacts, conductors, or grounding devices in various combinations.

known-good board: See *golden board.*

laminate: The plastic material, usually reinforced by glass or paper, that supports the copper cladding from which circuit traces are created.

laminate thickness: Thickness of the metal-clad base material, single- or double-sided, prior to any subsequent processing.

laminate void: An absence of epoxy resin in any cross-sectional area that should normally contain epoxy resin.

lands: The portions of the conductive pattern on printed circuits designated for the mounting or attachment of components. Also called *pads*.

layup: See *stackup*.

laser photoplotter: Creates an image of the individual object from a CAD database and then plots the image at very fine resolution.

lead-free HASL: A RoHS alternative to regular HASL where the physical application is the same, but the molten bath is free of lead and uses a nickel-modified alloy instead to give similar results.

leads: Projecting "legs" of components that pass through holes in the board and attach to lands on the other side. See also *lands*.

legend: A format of lettering or symbols on the printed circuit board. For instance, it may include the part number, serial number, component locations, and patterns.

lifted pad: A pad or land that has partially separated from its base material.

liquid photoimageable solder mask (LPI): A mask sprayed on using photographic imaging techniques to control deposition.

line: See *conductor*.

lot: A quantity of circuit boards that share a common design.

major defect: A defect that is likely to result in failure of a unit or product by materially reducing its usability for its intended purpose.

mask: A material applied to enable selective etching, plating, or the application of solder to a PCB. Also called *solder mask* or *resist*.

measling: Discrete white spots or crosses below the surface of the base laminate that reflect a separation of fibers in the glass cloth at the weave intersection.

metal foil: The plane of conductive material of a printed board from which circuits are formed. Metal foil is generally copper and is provided in sheets or rolls.

microsectioning: The preparation of a specimen of a material, or materials, that is to be used in metallographic examination. This usually consists of cutting out a cross-section, followed by encapsulation, polishing, etching, and staining.

microvia: Usually defined as a conductive hole with a diameter of 0.005 inches or less that connects layers of a multilayer PCB. Often used to refer to any small geometry connecting holes created by laser drilling.

minor defect: A defect that is not likely to result in the failure of a unit or product, or that does not reduce its usability for its intended purpose.

mounting hole: A hole used for mechanical mounting of a printed board (for instance, to the chassis), or for mechanical attachment of components to the board.

multilayer board (MLB): Printed boards consisting of a number (four or more) of separate conducting circuit planes separated by insulating materials and bonded together into relatively thin homogeneous constructions with internal and external connections to each level of the circuitry as needed.

NC drill (numeric control drill machine): A machine used to drill the holes in a printed board at exact locations, which are specified in a data file.

NC drill file: A text file that tells an NC drill where to drill its holes.

NC fabrication equipment: Numerically controlled machine tools such as routers and drilling machines.

net: An independent set of circuit nodes on a schematic that is connected to define an isolated circuit.

net list: An alphanumeric listing of symbols or parts and their connection points that are logically connected in each net of a circuit.

no-clean soldering: A soldering process that uses a specially formulated solder paste.

nomenclature: Identification symbols applied to the board by means of screen printing, inkjetting, or laser processes. See *legend*.

non-functional pad: A land on an internal or external layer that is not connected to an active conductive pattern on that layer.

non-recurring engineering cost (NRE): The one-time cost of design and development activities prior to starting production of a PCB.

non-wetting: A condition where a surface has contacted molten solder, but the solder has not adhered to all of the surface, so that some base metal remains exposed.

outer layer: The top and bottom sides of any type of circuit board.

pads: See *lands*.

panel: A rectangular sheet of base material or metal-clad material of predetermined size that is used for the processing of printed boards and, when required, one or more test coupons.

pattern: The configuration of conductive and nonconductive materials on a panel or printed board. Also, the circuit configuration on related tools, drawings, and masters.

pattern plating: The selective plating of a conductive pattern.

performance document: IPC specifications and guidelines specific to the PCB industry.

photographic image: An image in a photo mask or in an emulsion that is on a film or plate.

photoplotting: A photographic process whereby an image is generated by a controlled light beam that directly exposes a light-sensitive material.

photo print: The process of forming a circuit pattern image by hardening a photosensitive polymeric material by passing light through a photographic film.

phototool: A transparent film that contains the circuit pattern, which is represented by a series of lines of dots at a high resolution.

plated through-hole: A hole with plating on its walls that makes an electrical connection between conductive layers or external layers (or both) of a printed board.

platen: A flat plate of metal within the lamination press in between which stacks are placed during pressing.

plating void: The area of absence of a specific metal from a specific cross-sectional area.

plotting: The mechanical converting of X-Y positional information into a visual pattern such as artwork.

prepreg: Sheet material (for instance, glass fabric) impregnated with a resin cured to an intermediate stage (B-stage resin).

pressing: The process by which a combination of heat and pressure is applied to a "book," thereby producing fully cured laminated sheets.

printed board: The general term for completely processed printed circuit or printed wiring configurations. It includes single, double-sided, and multilayer boards, both rigid and flexible.

printed circuit: A conductive pattern that comprises printed components, printed wiring, or a combination thereof, all formed in a predetermined design and intended to be attached to a common base. In addition, this is a generic term used to describe a printed board produced by any number of techniques.

printed circuit board (PCB): A part manufactured from rigid base material upon which completely processed printed wiring has been formed.

pulse plating: A method of plating that uses pulses instead of a direct current.

reflow: The melting of an electrodeposited tin/lead, followed by solidification. The surface has the appearance and physical characteristics of being hot-dipped.

registration: The degree of conformity to the position of a pattern, or a portion thereof, a hole, or other feature to its intended position on a product.

resin (epoxy) smear: Resin transferred from the base material onto the surface of the conductive pattern in the wall of a drilled hole.

resist: Coating material used to mask or to protect selected areas of a pattern from the action of an etchant, solder, or plating. Also called *solder mask* or *mask*.

Restriction of Hazardous Substances Directive (RoHS): This directive restricts the use of six hazardous materials in the manufacture of various types of electronic and electrical equipment.

rigid-flex: A PCB construction combining flexible circuits and rigid multilayers, usually to provide a built-in connection, or to make a three-dimensional form that includes components.

rough holes: Holes with a copper burr around either the entry or exit hole and that lack a smooth barrel.

router: A machine that cuts away portions of the laminate to form the desired shape and size of the printed board.

schematic: A diagram, drawing, or plan that shows, by means of graphic symbols, the parts, electrical connections, and functions of a specific circuit arrangement.

scoring: A technique in which grooves are machined on opposite sides of a panel to a depth that permits individual boards to be separated from the panel after component assembly.

screen printing: A process for transferring an image to a surface by forcing suitable media through a stencil screen with a squeegee.

secondary side: The side of the assembly that is commonly referred to as the "solder side" in through-hole technology.

signal layer: An interconnection layer on a circuit board devoted exclusively to the routing of signal traces.

single-sided board: A printed board with conductive pattern on one side only.

SMOBC: See *solder mask over bare copper.*

solder: An alloy that melts at relatively low temperatures and is used to join or seal metals with higher melting points. A metal alloy with a melting temperature below 427 degrees C (800 degrees Fahrenheit).

solder leveling: The process by which the board is exposed to hot oil or hot air to remove any excess solder from holes and lands.

solder mask: Coating material used to mask or to protect selected areas of a pattern from the action of an etchant, solder, or plating. Also called *resist* or *mask.*

solder mask over bare copper (SMOBC): A method of fabricating a printed circuit board that results in final metallization being copper with no protective metal. The non-coated areas are coated by solder resist, exposing only the component terminal areas. This eliminates tin lead under the pads.

solder wicking: Capillary action, caused by surface tension, leads solder to fill small spaces such as holes, between strands of wire up a pad or component lead, or under the insulation of a covered wire.

stackup: The process in which treated prepregs and copper foils are assembled for pressing.

stencil: A metal sheet bearing a circuit pattern cut into the material.

step-and-repeat: A method by which successive exposures of a single image are made to produce a multiple image production master.

stripping: The process by which imaging material (resist) is chemically removed from a panel during fabrication.

substrate: A material on whose surface adhesive substance is spread for bonding or coating. Also, any material that provides a supporting surface for other materials used to support printed circuit patterns.

subtractive processing: The method of selectively removing copper from a board to form a circuit. In this case, "subtractive" refers to the method of image transfer from a phototool or image file to the copper circuit.

surface-mount technology (SMT): Defines the entire body of processes and components that create printed circuit board assemblies with leadless components.

Td (decomposition temperature): The temperature at which material weight changes by 5 percent. This parameter determines the thermal survivability of the resin material.

test coupon: A portion of a printed board or of a panel containing printed coupons used to determine the acceptability of such a board.

Tg (glass transition temperature): The temperature at which the resin system changes from a rigid or hard material to a soft or rubber-like material.

thieves (plating thieves also known as "robbers"): Non-functional metal areas on a surface to be electroplated. Their purpose is to balance the current density during plating to ensure uniform buildup of plated material.

through-hole: See *plated through-hole.*

time domain reflectometry (TDR): A technique that is used for measuring the characteristic impedance of a printed circuit trace.

tooling: See *non-recurring engineering cost.*

tooling holes: The general term for holes placed on a PCB or a panel of PCBs for registration and hold-down purposes during the manufacturing process.

top side: See *component side.*

trace: A common term for conductor.

traveler: The list of instructions describing the board, including any specific processing requirements. Also called a *shop traveler, routing sheet, job order,* or *production order.*

twist: A laminate defect in which deviation from planarity results in a twisted arc.

UL: Underwriters Laboratories, Inc., an independent product safety testing and certification organization.

UL symbol: A logotype denoting that a product has been recognized (accepted) by Underwriters Laboratories, Inc.

UV curing: Polymerizing, hardening, or cross-linking a low molecular weight resinous material in a wet coating ink using ultra violet light as an energy source.

via: A plated through-hole that is used as an interlayer connection but does not have component lead or other reinforcing material inserted in it.

void: The absence of any substances in a localized area.

warp (warpage): The deviation from flatness of a board characterized by an approximately cylindrical or spherical curvature. Also referred to as *bow* and *twist*.

Waste Electrical and Electronic Equipment Directive (WEEE): An international directive that sets collection, recycling, and recovery targets for electrical goods and is part of a legislative initiative to solve the problem of huge amounts of toxic e-waste.

wave soldering: A process wherein assembled printed boards are brought in contact with a continuously flowing and circulating mass of solder, typically in a bath.

wet solder mask: Applied by means of distributing wet epoxy ink through a silk screen.

wetting: Wetting in soldering applies to molten solder spreading along the base metal/metallisation surfaces to produce a relatively uniform, smooth, unbroken, and adherent film of solder. A good intermetallic bond between surfaces is formed.

whisker: A slender needle-shaped growth between conductors and lands that occurs after the printed board has been manufactured.

wicking: Absorption of liquid by capillary action along the fibers of the base metal.

work in progress (WIP): A list of orders that are in the manufacturing process and that have yet to be completed. Doesn't include finished product that is inventory.

X-raying: An inspection process used mainly for determining the alignment of internal features of a multilayer board.

About the Author

GREG PAPANDREW IS the co-founder of Better Board Buying and its subsidiaries, DirectPCB and PCBOffice. He's been saving circuit board customers money for over 25 years.

Greg pioneered a number of innovative approaches to offshore PCB purchasing, including high-mix, low-to-medium volume production. He was also among the first to deploy Asia-based personnel to successfully manage the manufacture of quality product delivered to the United States on time, at competitive prices.

Greg has long been steeped in every step of the board buying and manufacturing cycle. He has sold PCBs directly for various fabricators and, for over a decade, he ran one of the fastest-growing circuit board distributors in the United States. He has seen the industry grow and change. He knows what works and what does not work.

Now, he's once again upending traditional industry practices that have grown stale and bloated with unnecessary expenses. He is using his hard-won expertise to help circuit board buyers and to offer PCB customers a closer, more cost-effective relationship with overseas manufacturers.

For more information, visit BoardBuying.com.

Better Board Buying

B3 teaches buyers in the highly competitive PCB industry how to evaluate the strengths and weaknesses of both offshore and domestic manufacturing sources.

We also show them how to leverage buying power to get the best prices, and how to expertly navigate sourcing issues based on order size and technology type.

What You'll Learn from B3:

- How to better manage current vendors and assess new vendors.

- How to set service expectations for vendors and implement vendor-managed inventory programs that will meet changing production needs without raising costs.

- How to develop and implement corporate PCB fabrication specifications unique to your organization to ensure consistent quality.

- How to understand the differences between workmanship guidelines (IPC-A-600) and qualification/testing specifications (IPC-6012), and when to employ each.

- How to negotiate better terms and put into place rebate programs so vendors can reward your company's buying and payment performance.

- How to leverage your company's annual spend with carefully selected vendors, ensuring the best service while also diversifying your vendor base.

- How to move business from one vendor to another, when necessary, with minimal disruption to production schedules.

- How to navigate the strengths and weaknesses of both offshore and domestic manufacturing sources.

And a whole lot more...

Want more information? Ready to get your procurement team started in B3's workshop and certification program?

Contact Greg Papandrew at greg@boardbuying.com
or visit BoardBuying.com.

DirectPCB®

A B3 Company

Got Gerbers?® Cut Costs

You no longer need an expensive domestic broker to intercede with your off-shore vendors. DirectPCB® has a better way that will save you money while still delivering high-quality boards on time.

DirectPCB has already done the work of weeding out low performers and pre-qualifying top-tier overseas vendors. We get your orders into those vendors at the most competitive prices. And our offshore personnel conduct on-site expediting and inspection—while offering round-the-clock communication—so you always know where your boards are.

We do all this for less than the cost of an old-fashioned broker.

Why pay for an exorbitant U.S. middleman? It's an outdated model that's costing you money. DirectPCB® gives you the best of both worlds: offshore pricing and stellar, vendor-based service.

Want to learn more? Contact us at sales@directpcb.cn.

PCBOFFICE™

A B3 Company

 PCBOffice™

Supercharge your PCB Purchasing Power!

Want a more robust and cost-effective supply chain? Shrink it. Remove the expensive, U.S.-based middleman and save 10 to 25 percent on your PCB spending.

You can do that with PCBOffice.™ Our Premium Buying Package—available only from B3—is designed for companies with an annual circuit board budget of more than $250,000 who want the power of a full-scale overseas operation without its daunting costs and complications.

We strengthen your entire PCB buying cycle by placing your orders only with the top board makers. As a member of PCBOffice, you'll gain exclusive access to our Select Manufacturing Partners, a rigorously vetted group who will be highly responsive to your needs while consistently building quality product and delivering it on time.

For a low monthly fee, you'll get all of the on-site services provided by DirectPCB®—a team of experienced customer service agents, buying experts, engineers, and inspectors—but you'll get them at straight-from-the-vendor pricing. And that's just for starters. You'll also have your own bilingual PCB

Concierge who will be responsible for overseeing your projects from quote to delivery at your door.

Here's a detailed breakdown of what you'll get as a PCBOffice member:

- Straight-from-the-vendor pricing. It doesn't get more cost-effective than that.

- A bilingual PCB Concierge who lives near the supplier's facility. The Concierge will be your go-to person, ensuring your manufacturing, quality, and shipping guidelines are followed, and taking responsibility for your orders from quote to delivery at your door. Your Concierge will also facilitate your corporate or customer visits.

- Job tracking with daily WIP and SHIP reports.

- CAM, engineering assistance, and EQ resolution.

- Data analysis of quote, quality, and delivery performance among your Select Manufacturing Partner team.

- On-site expediting, source inspection, and swift quality problem resolution.

- Warehousing and distribution services management.

Having your own offshore PCB office will supercharge your purchasing power, giving you a leg up on the competition. Your customers will think you're a super hero. And you'll sleep soundly at night, knowing your boards are in the best hands.

Want to find out more about the groundbreaking PCBOffice™ model of circuit board purchasing? Visit BoardBuying.com.